Quarto.com

First Published in 2026 by Cool Springs Press, an imprint of The Quarto Group,
100 Cummings Center, Suite 265-D, Beverly, MA 01915, USA.
T (978) 282-9590 F (978) 283-2742

EEA Representation, WTS Tax d.o.o.,
Žanova ulica 3, 4000 Kranj, Slovenia.
www.wts-tax.si

Cool Springs Press titles are also available at discount for retail, wholesale, promotional, and bulk purchase. For details, contact the Special Sales Manager by email at specialsales@quarto.com or by mail at The Quarto Group, Attn: Special Sales Manager, 100 Cummings Center, Suite 265-D, Beverly, MA 01915, USA.

30 29 28 27 26 1 2 3 4 5

ISBN: 978-0-7603-9889-0

Digital edition published in 2026
eISBN: 978-0-7603-9890-6

Library of Congress Cataloging-in-Publication Data is available.

Design: Brunch Design Company
Cover Image: Shutterstock
Page Layout: Sporto
Photography: Frank Hyman, except pages 6, 18, 48, 62, 84, 106, 132, 154, 168, 182, and 204 (Shutterstock)
Illustration: Mattie Wells Design

Printed in Guangdong, China TT112025

RIPE TOMATO REVOLUTION

Planting and Growing Every Type of Tomato: Beefsteaks, Cherries, Plums, Dwarfs, and Heirlooms

FRANK HYMAN
author of *How to Forage for Mushrooms Without Dying*

COOL SPRINGS PRESS

CONTENTS

CHAPTER 1

Join the Revolution

IF A FIRST-TIME GARDENER GROWS ONLY ONE VEGETABLE, IT WILL BE A TOMATO. AND THEY WILL *LOVE* THE FLAVOR OF THAT HOME-GROWN VEGETABLE. (OR IS IT A FRUIT? DON'T WORRY, WE'LL GET TO THAT.)

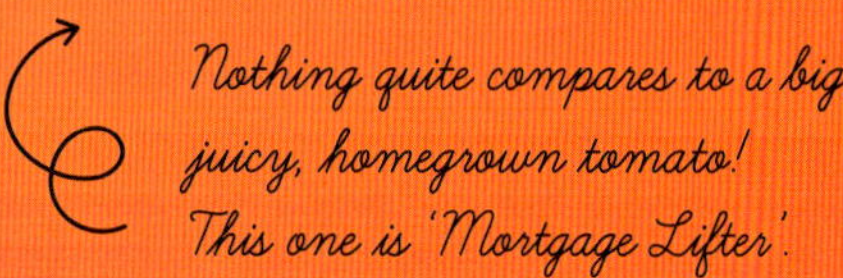

Nothing quite compares to a big, juicy, homegrown tomato! This one is 'Mortgage Lifter'.

But they won't love the trials caused by hungry bugs, rampant diseases, feisty weeds, and floppy tomato plants that want to be tied up like so many prisoners.

But it doesn't have to be that way. That's why I wrote this book.

I've grown backyard tomatoes for four decades. Which is another way of saying I've already grown them all the wrong ways so that you won't have to.

Also, by training, I'm a horticulturist. But by nature, I'm a *how-to*-culturist, always looking to get the most sustainable, best-looking, most cost-effective results with the least botheration.

This book isn't just aimed at novice tomato growers. I've honed my practices so that even intermediate and advanced gardeners will learn things they didn't know. And perhaps even more important, maybe they'll *unlearn* a few things that were never true in the first place.

So, what do I mean by "revolution"? Well, I don't mean we're going to storm the Bastille and capture the queen. I mean, the tomato plant *is* the Queen (of the garden, at any rate). It's something akin to the Print Revolution or the Scientific Revolution or the Punk Music Revolution. Not on the same scale, obviously, but of a similar nature. (And perhaps with less hair dye. Certainly with fewer powdered wigs.) What I mean is a global revolution in the way of thinking about tomato gardening that yields a net benefit for all. In this book I'll be sharing these six revolutionary tomato garden options that I've not seen in any book, blog, or magazine (unless I wrote it).

1. Tomato Houses (chapter 5).
2. Never-stinks-and-never-needs-turning compost (chapter 10).
3. Mobile compost bins (chapter 10).
4. Baby greenhouses (chapter 8).
5. Horse trough water garden/rain barrel (chapter 7).
6. Push water uphill with a rain barrel (chapter 7).

And six more options that deserve wider use:

7. Fertilizer from firewood (chapter 3).
8. Straw bales as a snack patio (chapter 2).
9. Coffee chaff and coffee bag mulch (chapter 9).
10. Two kinds of raised beds that give the garden a finished look (chapter 2).
11. Three kinds of time-saving DIY tomato trellises (chapter 5).
12. DIY self-watering pots (okay, I got this one from my buddy Steve Biggs at foodgardenlife.com).

An ephemeral sculpture I call, "And a Cherry on Top."

Growing tomatoes *does not* **have to be** *so hard.*

Tomato Truth versus Tomato Fiction

I tell my garden clients that the best design comes from a dialogue between us. Likewise, this book will be best understood as a conversation between you and me. Granted, I can't hear you from my house. But you are invited to talk back.

What do you know about the world's most favorite homegrown veggie? We're about to find out! This tomato is an heirloom called 'Gold Medal'.

I encourage you to use this book in the way I use all the books in my library: I consider them tools for living my life. As such, I'm allowed to fill the margins with notes—even in ink! Plus, I give myself permission to dog-ear pages, to fire off asterisks and exclamation points, and to underline passages.

I do these things knowing I'll come back to these books in time and cut straight to the highlights. I want you to be able to do the same with this book. I want you to open this book in a few years and be able to quickly find the passages that help you navigate the problems—or opportunities—at hand most effectively.

Also, when it comes to nonfiction, I often don't read chapters in order, the way one would a novel. So, feel free to scope out the table of contents and the index and choose your own adventure. Doing so will keep our conversation at a high level of engagement, even if you don't read the book in a linear fashion.

TO HELP US START OUR CONVERSATION, I'm offering some notions about growing tomatoes on the following page. It's a quiz of sorts (but a fun one!). Feel free to mark up this page of true or false propositions. Get your hands dirty. Give your copy of this book its own character.

True *or* False

1.
Tomatoes are poisonous.

☐ TRUE ☐ FALSE

2.
Tomatoes are annuals.

☐ TRUE ☐ FALSE

3.
Tomatoes are vines.

☐ TRUE ☐ FALSE

4.
Tomatoes are fruits.

☐ TRUE ☐ FALSE

5.
Professional plant breeders created cherry tomatoes.

☐ TRUE ☐ FALSE

6.
Tomatoes are the queen of the vegetable garden.

☐ TRUE ☐ FALSE

7.
As a kid, the author hated eating tomatoes.

☐ TRUE ☐ FALSE

8.
Tomatoes are super easy to grow.

☐ TRUE ☐ FALSE

9.
This book can help you grow more and better tomatoes.

☐ TRUE ☐ FALSE

True statements: 3, 4, 6, 7, 9 - False statements: 1, 2, 5. - "It depends" statements: 8

Curious about *why* your tomato-y answers were correct or incorrect? *Here's the scoop.*

1. FALSE.

Some people in northern Europe and North America thought tomatoes were poisonous because they were in the nightshade family, which does include poisonous plants such as belladonna and tobacco. This misunderstanding about the edibility of the tomato and the fact that they were associated with witches and were-wolves, led the Swedish botanist Carl Linnaeus, to choose the Latin name *Solanum lycopersicum*, which translates to "Nightshade Wolf Peach."

2. FALSE.

Trick question. Tomatoes are sold and grown as annual vegetables because they won't survive even a mild winter in cold regions. In truth, they are perennials that live for a few years in their home terrain of Peru and northern Chile where the winters are warm, and the rainfall is modest (more about this in the sister regions section on p. 160).

3. TRUE.

Tomatoes are vines, but they do not have climbing tendrils or a twining growth habit like peas or pole beans. Unattended, tomato vines ramble along the ground somewhat like a ground cover. However, tomatoes share the general ability of vines to easily establish roots from any part of their stem.

4. TRUE.

As one anonymous person said, *intelligence* is knowing that a tomato is a fruit because it has seeds, but *wisdom* is knowing that because the flavor is savory, you don't add it to a fruit salad.

5. FALSE.

It's the other way around. The original tomatoes were the size of cherry tomatoes. Breeders created bigger varieties over the centuries.

6. TRUE.

No need to say more.

7. TRUE, SADLY.

I was a teenager before my taste buds found tomatoes acceptable. Unless they were on a pizza.

8. IT DEPENDS.

It depends on how much rainfall your region gets. And on evolution. More details in the sister regions section on page 160.

9. TRUE!

Let's get started!

Why are tomatoes **so difficult to grow in so many regions? Because** *that's what happens when* **evolution meets globalization.**

The Trouble with Tomatoes

The biggest challenges with growing tomatoes in most parts of the world are their tendencies to get root rot and leaf maladies. It makes you wonder how a plant could possibly evolve with such a poor immune system. That's a backward way of looking at the problem, however. I did find a good answer to that conundrum, but I had to do some international horticultural sleuthing to find it.

The beginning of my understanding of this trouble with tomatoes goes back to spending two gorgeous spring weeks on my own in the boondocks of Morocco back in the double aughts. On a trail in the treeless Atlas Mountains I befriended a twenty-two-year-old goat herder in a yellow turban named Yusef. He spoke French about as badly as I did. But we hit it off, and I understood enough when he said, "You could come home with me, meet my family, help me slaughter a goat for dinner, and spend the night in our cave."

A tomato sandwich on "no-knead" bread made by my wife, Chris Crochetiere.

The horticultural lesson came from the numerous spring-fed organic farms and gardens we passed on the way to his family cave, many of them with a crop of tomatoes. Organic growing sites seemed more common in Morocco than they were in the United States or Europe. Why? Sometimes what scientists call "ground truth"—which means walking the actual territory—reveals a lot. With much lower rainfall, the gardens, farms, and orchards there contend with significantly fewer insects, diseases, and weeds. So, growing organically was much easier in that drier climate. Ahh, enlightenment.

Returning to the States, I looked at my tomato plants that summer and noted their fuzzy leaves. I'd always known them to be fuzzy, but I hadn't given it a thought before. Generally, hairy plant leaves are an indication of having evolved in a dry, sunny climate, which I now thought of as the same conditions that inhibit pests. If you think of other plants with fuzzy leaves—lamb's ear, rose campion, lavender, figs—you'll find they evolved in dry, sunny regions, like the Mediterranean.

Most gardeners know that the first tomatoes grew thousands of years ago in coastal Peru and northern Chile. If those original tomato plants were as disease-prone in that area as they are for most gardeners around the globe, evolution would have weeded tomatoes out long ago.

But a rainfall map of South America reveals that as the winds push rain clouds west over the Amazon jungle and up the Andes, they drop their moisture as they rise higher. Once the clouds have crossed the peaks of the Andes, the west side of this mountain range becomes what's called a "rain shadow." Meaning the Pacific side is drier than the Atlantic-facing side of the Andes. In that thin strip of land, the annual rainfall levels are between 10 and 20 inches.

Huh. That's similar to the annual rainfall in Mediterranean countries like Spain and Italy. So plants that evolve in the rain shadow of the Andes are subjected to fewer pests and don't develop much of an immune system for diseases.

The Aztecs of what is now called Mexico had extensive trading paths into South America. No doubt, some of their traders brought tomato seeds home and introduced them as a new food to their empire.

When the Spanish invaded the Aztec territory, they found tomatoes growing in what's now called the Vera Cruz region.

Here, in doing my research, I was almost thrown for a loop. The average precipitation in Vera Cruz in summer runs as high as ½ inch (1.25 cm) every day, or about 60 inches (1.52 m) per year. So much for my theory about tomatoes doing well in dry conditions.

A closer look at the rainfall chart for Vera Cruz, however, reveals that almost all that rain comes in summer. Winter is pretty dry and plenty warm enough to grow tomatoes. Here in the States, when we get tomatoes from Mexico, they arrive in winter.

So, this means the Aztecs weren't dealing with the kinds of tomato diseases that many of us face now because their winter season wasn't really cold, they only received about 10 inches (25.4 cm) of rain in winter, and the Aztecs' engineering was sophisticated enough—think Romanesque dams and aqueducts—to provide supplemental irrigation that didn't wet the foliage and cause leaf maladies or make the soil soggy enough to cause root rot.

And so, in the 1500s, the Spanish colonizers brought tomatoes home with them, to places where—aside from the rainy northwest coast—there was an annual rainfall of about 20 inches (50.8 cm). Tomato plants thrived under those conditions and captured the affection of the Spanish and other Mediterranean countries with modest annual rainfall, such as Italy. Cue the pizza and spaghetti sauce. Read more about what are called sister regions with similar weather in chapter 8 on page 163.

HOW ABOUT AN ALTERNATIVE HISTORY?

What might have happened if the first European conquerors of the Aztecs in the 1500s had been the English instead of the Spanish?

If the Brits had been the ones to bring home tomatoes, it might have taken much longer for tomatoes to gain a fan base in Europe. The steady, drizzly rainfall of the British Isles would have yielded mostly diseased and dead tomato plants.

As it was, even with a Spanish introduction to these marvelous fruits, tomatoes didn't become common in rainy England until the 1800s, when sheets of glass for greenhouses became available to aristocrats. With a way to keep raindrops from splashing on the leaves and by limiting soil moisture, the English finally came to see the virtue of this plant.

By then, breeders had begun developing varieties that could struggle through a summer season in wet places, like the United States, eastern Asia, and other parts of the world.

And that brings us to today, where, in the rainy parts of the world, gardeners have to be on their game to keep alive plants with a modest immune system that evolved in a drier climate.

Tomatoes: The Savory Fruit

Given its history, it seems a minor miracle that we eat as many tomatoes as we do. For some in the 1600s, tomatoes were associated with werewolves. In the next century, Linnaeus chose the Latin name *Solanum lycopersicum*, which translates to "nightshade wolf peach."

Some English emigrants in America rejected tomatoes all the way up to 1860. The US Civil War finally mainstreamed tomato eating. This particular aversion to a healthy food item gave rise to the medical shorthand known as "the tomato effect," a term for effective therapies avoided for cultural reasons.

Thank goodness all that's behind us, and we can now look forward to multiple medleys of 'maters. So, which ones will you plant this year? And why? Here are some of my choices for top tomato.

Big, slicing tomatoes for grilled cheese sandwiches and bright salads can be had by growing locally popular varieties like 'Cherokee Purple', 'German Johnson', 'Gold Medal', and 'Brandywine'. These sturdy heirloom varieties produce all season and grow very tall. These types of tomato are known as "*in*determinates." That means they keep fruiting and getting taller until frost kills them. The best way to keep indeterminate plants off the ground is with a 5-foot (1.52 m)-tall tomato cage made from a heavy roll of remesh sourced from a big-box building supply store (see p. 88).

Tomatoes for sauces—like 'Roma', 'San Marino', and 'Amish Paste'—are determinate varieties that grow to only about 4 feet (1.21 m) tall. They also finish their season early with lots of ripe paste tomatoes that ripen at the same time. You can plant a second crop in very late spring or early summer to extend the season.

Starting with a mix of plum tomatoes, heirlooms, dwarfs, and hybrids will help you learn which ones taste best to you and which ones do well in your particular conditions. Or at least in this year's conditions.

Cherry tomatoes that are popular include 'Sungold' and 'Sweet 100'. Both are indeterminate and will need tall trellises. Shorter cherry varieties, such as 'Super Sweet 100' and 'Tiny Tim', grow only 2 to 3 feet (60.96 to 91.44 cm) tall.

New dwarf tomatoes like 'Rosella Purple', 'Dwarf Sweet Sue', and 'Mr. Snow' grow to 3 to 4 feet (91.44 cm to 1.21 m) and can survive in large pots (5-gallon, or 18.92 L, buckets or similar) and in straw bales.

In most cases and in most places, cherry tomatoes have fewer problems. And more opportunities for snacking and sharing.

LET'S *do a* DEEPER DIVE

Advice from an innovative gardener who was also an actual political revolutionary couldn't hurt, right? Thomas Jefferson largely wrote the Declaration of Independence and also wrote enough notes about his gardening practices—including growing tomatoes for eating in the 1780s—at his homes Monticello and Poplar Forest to create a book called *Thomas Jefferson's Garden Book.*

Now that you know tomatoes are from a region with a climate similar to the Mediterranean, you'll learn a lot from Peter Dallman's *Plant Life in the World's Mediterranean Climates.* For example, the five regions with a Mediterranean climate comprise only 2 percent of the world's land area, but they boast 20 percent of the world's plant species.

And English gardener Beth Chatto's *The Dry Garden* addresses the various "rain shadows" that form in drier parts of wet regions, like the British Isles and America's Pacific Northwest. These microregions allow gardeners to grow Mediterranean-type plants, such as tomatoes, without the rampant diseases that wet regions normally suffer from. In her book, Chatto laments her inability to grow astilbes and other "English" flowers. But after visiting the Mediterranean island of Corsica, she learned to introduce plants that favor a dry summer and her garden took off.

Quick List *of* Best Practices *for* Any Kind *of* Tomato

- Because sunlight is food for plants, give tomatoes full sun, or at least 6 to 8 hours of direct sun. More sunlight = more plant food = more and bigger tomatoes.

- Apply lime to the soil to make nutrients more available and to minimize blossom end rot (see p. 110).

- Set out transplants only when the ground is warm enough and after the danger of frost has passed. Depending on your climate, that can be anytime from March through May. For most varieties, expect ripe fruit about 3 months later.

- Dig a hole for your tomato transplants and drop two big fistfuls of organic fertilizer into the hole.

- Remove the lower tiers of leaves before planting and keep removing the lowest leaves as the plant grows. Otherwise, diseases in the soil will be bounced by the rain up onto the lower leaves. As the plant grows, keep a 1 to 2-foot (30.48 to 60.96 cm)-high gap—like a vertical moat—between the mulch and the lowest leaves.

- Don't irrigate with an overhead sprinkler. Wet leaves just promote diseases. Use a soaker hose on the ground for irrigation and put a timer on the spigot. In summer, aim for 2 gallons (7.57 L) per square foot (0.09 sq. m) of garden bed per application. You'll learn more about irrigation in chapter 7.

- Cover the soil and the soaker hose with 2 inches (5 cm) of organic mulch: shredded leaves, pine needles or coffee chaff. In chapter 9, I discuss how mulch is like birth control for weeds, and we take a much deeper dive into the subject. Plus, mulch keeps the soil from drying out during droughts and from getting compacted by the rain.

- Trellis the tall growers—*in*determinates—with a trellising system. Chapter 5 introduces several options.

- Trellis shorter-growing determinates with what I call a cat's cradle (sometimes called a "Florida weave"), a technique introduced on page 86.

- Remove any leaves that show disease and trash them. Smoosh any caterpillars you see. An organic pesticide called Bt (*Bacillus thuringiensis*) will also control caterpillars. Pests and diseases are covered in depth in chapter 6.

- Studies show that removing suckers doesn't help in any way. If you don't know what suckers are, then just be glad.

- Tomatoes don't have to fully ripen on the vine to become delicious. They can be harvested after they've started to turn red and allowed to ripen in the kitchen, but out of the sun.

- Never put tomatoes in the fridge.

- Always eat tomatoes at every meal.

- Repeat next year.

Homemade tomato bruschetta served on slices of fresh baguette.

CHAPTER 2

Beds for Tomatoes: Raised and Otherwise

THE CONCEPT OF USING RAISED BEDS FOR GROWING VEGETABLES DEVELOPED IN NORTHERN EUROPEAN REGIONS WITH HIGH RAINFALL IN SPRING. RAISED BEDS DRAINED FASTER AND BABY VEGETABLES COULD GET MORE OXYGEN FROM THE SOIL THAN PLANTS GROWN IN GROUND-LEVEL BEDS. BUT THERE'S A TRADE-OFF IN SUMMER. RAISED BEDS IN DRY AREAS—OR IN DRY SEASONS LIKE SUMMER—CAN LEAVE YOUR VEGETABLES VERY THIRSTY. SO, INVEST IN MULCH, SOAKER HOSES, AND A TIMER TO MAKE SURE YOUR SUMMER VEGETABLE GARDEN IS GROWING EVERY DAY (ALL THINGS YOU'LL LEARN HOW TO DO IN THIS BOOK).

A good general rule is that a raised bed in summer will need about 1 gallon (3.78 L) of water, per square foot (0.09 sq. m), per week if it doesn't rain. This means that a 4 × 12-foot (1.21 × 3.65 m) raised bed will need about 48 gallons (181.7 L) of water per week. If you're not prepared to water quite so much, consider another type of bed for growing your tomatoes. In this chapter, we examine the many places and ways you can grow tomatoes—from raised beds to subirrigated planters and straw bales—and discuss the techniques and materials needed for each to help inform your choice.

Raised Beds: Better Drainage and a Tidier-Looking Garden

Vegetables need warm soil and good drainage, especially in spring. Raised beds are a classic tool for getting both.

Here's a quick rundown of how to make raised beds that look good and last a long time:

- Cedar planks look better than pressure-treated wood and last longer than untreated pine or oak.
- Seven-inch (17.78 cm)-tall beds (using 2 × 8s, or 38 ×184 mm, which are really only 7.25 inches, or 18.41 cm, wide) are tall enough to create a stronger visual impression than 2 × 6s (38 × 140 mm) or 2 × 4s (38 × 90 mm). Sixteen-inch (40.64 cm)-tall beds—stack a pair of 2 × 8s (38 ×184 mm) and cap them with a flat 2 × 6 (38 × 140 mm)—make a perfect height for sitting while one works (check the seat height of your favorite chair and it's probably 16 or 17 inches, or 40.6 or 43.1 cm). The illustrations and instructions on pages 27 through 31 show you how to make both a single- and double-height raised bed.
- Mitered corners (picture frames have mitered corners cut at a 45-degree angle) look better because there is no exposed end grain.
- Screws work better than nails: You can dismantle the beds if you ever move; screws require less physical strength to install than driving nails; and screws help boards resist twisting better than nails do.

Three seating-height raised beds with their seat caps installed after they've been filled with soil. The nearest bed will be filled with soil before *installing the seating cap. Note the scrap pieces of wood keeping the top and bottom planks aligned, and the landscape fabric covering the bottom to keep weeds out.*

The red color of heartwood indicates that it's rot resistant. The yellowish color of the sapwood indicates that it is not rot resistant. Be sure to specify that you want only heartwood for outdoor projects.

Single-height raised beds of heartwood cedar in our sunny front yard. It's a small yard, so I cut off the inside corners of each bed so the center wouldn't feel congested. And because the four beds are near the street, I blocked off three of the gaps between beds, so they would be less inviting to passers-by.

Seating-height raised beds at Chris' family place in Maine. Made from the heartwood of locally harvested white cedar. The 2 × 6 (38 × 140 mm) seating caps have been stained red for a delightful visual contrast.

The view of our front yard raised beds from the roof of our house.

A simple tool called a "speed square" helps assure a square corner and allows marking of perpendicular lines.

RAISING THE ISSUE OF PRESSURE-TREATED LUMBER IN RAISED BEDS

I've read at least half a dozen articles and blog posts, in as many years, warning gardeners not to use pressure-treated lumber because it contains toxic arsenic and chromium. It's disappointing that these "experts" don't know that, since 2003, those two chemicals haven't been used in residential lumber (meaning lumber sold at big-box stores and most lumberyards).

Pressure-treated lumber for raised beds, decks, and fences is still for sale, and it's still rot resistant, but it contains two very mild chemicals. These modern chemicals are as effective at keeping fungus and termites out of the wood as arsenic and chromium ever were, but they are very much safer—safe enough, in fact, to use in wood for raised beds for growing edible plants.

One replacement is a liquid form of copper, which many of you know is the same material from which modern water pipes are made. The water you drink, cook with, and shower in passes through copper water pipes every day. The other chemical is the same compound used in 409 spray cleaner (once owned by Art Linkletter, the 1940s, '50s, and '60s TV host, just in case trivia is your thing). It's been commonly used in homes on kitchen countertops around food and children since the 1960s. When sprayed in the air, some people can be sensitive to it, but in lumber, it's bonded with the wood and doesn't become airborne. Any that leaches into the soil would not travel far and would be very dilute. There are probably other products in your kitchen and bathroom that are more likely to harm you than modern pressure-treated lumber used as raised beds and fence posts in the garden.

That said, if you are certified as an organic grower, check with your certifying agency as there is some variation regarding what's deemed acceptable and what can be grandfathered in depending on the organization.

So, what kind of wood is used for making pressure-treated lumber? Because treating wood with chemicals costs money, the lumberyards use less expensive wood to make it. That means, in the eastern United States, southern yellow pine is used. It's heavier, denser, and more likely to twist than other more expensive pines. So, it's harder to drive a nail into. And you may need to use more nails (or for better holding power, screws) to keep your raised beds from twisting. On the West Coast of the United States, pressure-treated wood is commonly made from hemlock for the same reason. It's a less desirable wood than the lighter and straighter west coast SPF (spruce, pine, fir) that is commonly used for framing houses. But, once it's been treated, it's more expensive. Other countries have their own wood sources and preferences for pressure treating.

The benefit is that pressure-treated lumber will last as long as twenty years in contact with the soil. On page 26, I share some good non-pressure-treated lumber options for raised beds, if you are looking for alternatives.

In any event, if you come across an "expert" telling you to avoid treated lumber, let them know there isn't any arsenic or chromium in pressure-treated lumber anymore.

It's not necessary, but using a router with an ogee bit gives the top, outer edges of the beds an appealing architectural detail. And with mitered corners, you're never looking at the less attractive end grain of each board.

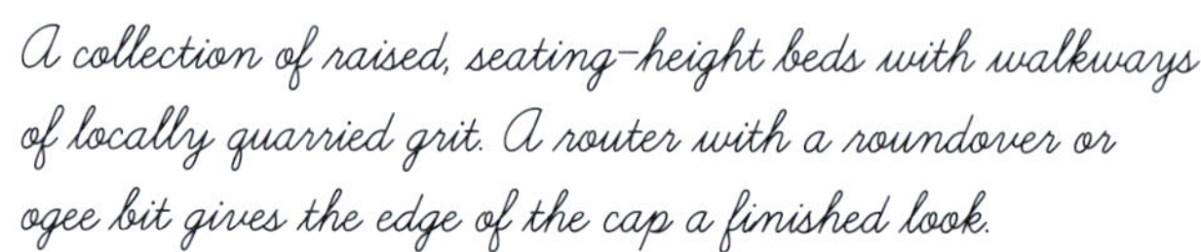

A collection of raised, seating-height beds with walkways of locally quarried grit. A router with a roundover or ogee bit gives the edge of the cap a finished look.

Time-Saving Tips *for* Filling Raised Beds *with* Topsoil

- Lay out a sturdy brown plastic tarp where you want the topsoil delivered. A 12 × 16-foot (3.65 × 4.87 m) tarp is enough for a 10-cubic yard (7.64 cubic m) load. The tarp will help the driver know where you want it and will greatly reduce the amount of time required to spread and clean up the last of the topsoil.
- Use long-handled shovels to load the topsoil into wheelbarrows.
- But don't poke holes in the tarp. Leave a 3-ish-inch (7.5-ish cm) layer of soil on the tarp as you work so the shovel doesn't tear the tarp. When you've moved most of the topsoil, lift the corners of the tarp to flip the topsoil into a higher pile. When there's only a small amount of soil left, lift the tarp and dump the last of the topsoil into a wheelbarrow. The tarp will last much longer without you poking holes in it.
- Your raised beds' walls will already be in place. Landscape fabric under the beds will keep out Bermuda grass and other weeds. Be sure not to disturb the fabric.
- Push the front of the loaded wheelbarrow up against the wall of the raised bed. Tip the wheelbarrow up so all the topsoil flows into the beds. It's okay to rest the front of the wheelbarrow on top of the raised bed wall.
- Use a gravel rake to spread the soil throughout the bed.
- Each bed should be filled to the top, which may require shoveling out the last few wheelbarrow loads instead of dumping them. No need to pack down the soil, the rain will do that for you.
- If you plan to install a flat seating cap on your bed's upper edge, install it after the soil has been placed.

Often it's easier to get all the soil out of the wheelbarrow by turning it upside down, rather than wagging it left and right on its nose.

RAISED EXPECTATIONS FOR RAISED BEDS

One of the first things I learned about home gardening was the importance of growing in a raised bed. As I helped my old college buddy James Forrest White shape up some beds, he taught me the primary reason for gardening this way. In a cold, wet spring, the excess rain drains out from the raised bed more quickly, sparing the vegetables growing in it from drowning.

Farmers in rainy climates—like northwestern Europe, a sister region to the rainy Pacific Northwest in the United States—first used raised beds. But raised beds aren't the answer for every gardener. In dry climates, like in the southwestern United States, for instance, Indigenous Native American farmers planted vegetables in beds slightly dug out *below* the natural soil level. That way, the beds would gather more of the rare rainwater and reduce the need for irrigation.

In climates that aren't as rainy as northwestern Europe and the Pacific Northwest, such as the plains, Midwest, and eastern United States, raised beds are beneficial during cool, rainy months and during summer downpours. But during summer dry spells, that quick drainage demands a bit more attention to irrigation and mulching to keep soil moisture levels steady.

But let's be honest and admit that part of the reason we have raised beds is because they can add structure and beauty to the garden. Having a neat edge gives us a sense of the garden being under control, even if the weeds are getting away from us.

However, it isn't absolutely necessary to define a raised bed with lumber. Simply tossing soil from the pathways onto the beds will raise them up enough to make a difference. The first raised beds I made for myself were that freeform style. But since they were about 2 feet (60.96 cm) wide and 6 feet (1.82 m) long, they did resemble a cemetery's burial plots.

Planks of rot-resistant eastern red cedar heartwood from a local mill will be used to rebuild our very old raised beds in the front yard.

So, ever since I acquired basic carpentry skills, I've built raised beds with a more permanent and less funereal look. And you can too.

TYPES OF WOOD

You have many wood choices for building raised beds, but if the wood isn't rot resistant, termites and fungi will slowly dismantle your beds for you. In about five years (sooner with high rainfall, later in a dry climate) you'll be starting over and rebuilding your beds.

You could also build them with wood that is naturally rot resistant. Here in the United States, untreated alternative rot-resistant woods for raised beds include redwood, black locust, cypress, white cedar, and eastern or western red cedar. These woods often cost more than pressure-treated wood, but generally will last only about half as long. And, even that comes with a caveat. Only the dark-colored, dense heartwood of these trees will resist rot. The heartwood could last as long as twenty years or more. But if the lumber is cut from the tree's sapwood, the boards won't last much longer than salvaged untreated pine. Why? The heartwood is the center of older trees and the cells of the heartwood are all dead. The live cells of the sapwood (between the bark and the heartwood) send all their waste material to be dumped in the dead cells of the heartwood. So, yes, that lovely smell of heartwood cedar is the "poop" from the sapwood of cedar. Who knew? Make sure you're buying solid planks of heartwood unless you want to rebuild your raised beds twice a decade.

However, these rot-resistant heartwood planks aren't readily available everywhere. You may need to find a local lumberyard or high-end lumber supplier. And where they are available, they can be pricey.

RAISED BED SIZES

The standard width of a raised bed is 4 feet (1.21 m). Any wider than that and the average gardener may not be able to stand in the path and reach the center of the bed. But 3 feet (91.44 cm) wide is fine for gardeners without as much reach. Most often, the length of the bed is determined by the length of planks that can be bought or handled safely. The smallest bed I recommend is a square one made from four 4-foot (1.21 m)-long 2 × 4s (38 × 90 mm).

The size I recommend most often are beds made of 2 × 8s (38 ×184 mm) that are 4 feet (1.21 m) wide and 8 feet (2.43 m) long. That way, each bed can be made from three 8-foot (2.43 m)-long 2 × 8s (38 ×184 mm). With one 8-footer (2.43 m) cut in half, you'll have the two 4-footers (1.21 m) you need for each end of the bed. I like this size because the proportions look right, the lumber fits even in a small pickup truck, you can grow a lot of vegetables in 32 square feet (2.97 sq. m), and the depth allows you to add lots of topsoil and/or compost and mulch while keeping a neat edge.

TOOLS FOR BUILDING RAISED BEDS

Basic carpentry tools will get the job done.

A corded or cordless drill with an assortment of drill bits and driver bits lets you drill pilot holes and drive 3-inch (7.62 cm) exterior-grade screws into place. As I've already mentioned, screws are better than nails: You don't risk smashing your thumb; the boards don't pull apart over time; and you can take apart the bed if you want to move it.

A circular saw with a new blade will cut everything to length. But they can be loud and scary, so get some tutoring from an experienced carpenter before using one.

Tools for measuring and marking are also important: a tape measure, a pencil, and a speed square for marking cut lines.

You'll also want safety gear, like leather gloves to protect your hands from splinters as well as ear and eye protection to save your hearing and eyesight.

Cords, sawhorses, and clamps round out the simple tools you'll need for this project.

Last, you'll want some open space for the sawhorses and a flat surface, like a patio or driveway, for screwing the sides of the beds together.

STEPS FOR BUILDING A BASIC RAISED BED WITH MITERED CORNERS

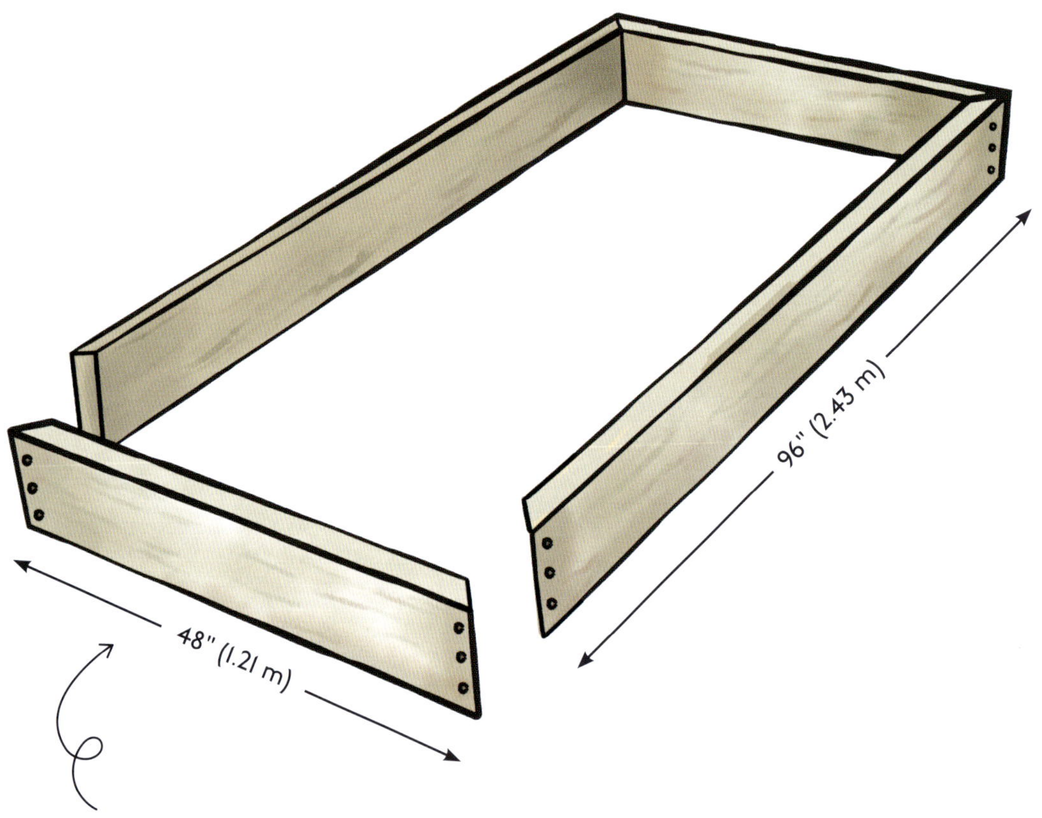

A basic plan for building a 4 × 8-foot (1.21 × 2.43 m) raised bed with mitered corners. Mostly it doesn't happen, but if one screw hits another during installation, just back it out, move its location slightly, then drive it in again to avoid the other screws.

1. Mark and cut the very ends of a pair of 8-foot (2.43 m) 2 × 8s (38 ×184 mm) with a 45-degree bevel. These cuts are harder to make than straight ones—so take your time. And clamp your planks to the sawhorse so they don't shift around during the process.

2. Mark and cut the opposite ends with a 45-degree bevel that's 96 inches (2.43 m) from the first cut. Or "long point to long point" as carpenters say.

3. Take the third 8-foot (2.43 m) 2 × 8 (38 ×184 mm) and cut one end with a 45-degree bevel.

4. From the bevel, mark and cut another 45-degree bevel 48 inches (1.21 m) away from the first: long point to long point.

5. Flip the leftover piece of that third board. It already has a 45-degree bevel cut at one end. So repeat Step 4.

6. Measure, mark, and cut a final 45-degree bevel that's 48 inches (1.21 m) away from the first one.

7. Install a drill bit that's slightly wider than the threads of your 3-inch (7.62 cm)-long screws in your drill.

8. Drill three pilot holes at the end of each board (twenty-four holes total), through the beveled part. The locations don't have to be precise as long as they are about 1 inch (2.54 cm) from all three edges. It's better if they aren't precise—that way the screws on one side of a corner won't conflict with the screws on the other side of the corner.

9. Put two corners together on a flat surface, like a patio or driveway.

10. Place a scrap of wood (ideally a 2 × 6, or 38 × 140 mm, or 2 × 8, or 38 × 184 mm, about 30 inches, or 76.2 cm, long) across the corner and clamp the scrap to the work pieces. Use a speed square to make sure you've got a 90-degree corner (as shown in the photo on p. 22).

11. Place six screws in the six holes. One by one, drill them into place. The scrap plank and clamps should keep the corner aligned.

12. Repeat steps 9, 10, and 11 on the other three corners.

13. Recruit someone to help you carry the finished raised bed into the garden. If conditions allow, you may be able to lift one end and drag it in place by yourself. You can also step into the center of the bed and hold it like a hula hoop around yourself.

14. Fill the raised bed with topsoil. A 4 × 8-foot (1.21 × 2.43 m) bed built with 2 × 8s (38 ×184 mm) will hold about 20 cubic feet (0.56 cubic m) of topsoil (a cubic yard is 27 cubic feet, or 0.76 cubic m).

15. Plant it up!

I'LL RAISE THAT RAISED BED!

For gardeners who have trouble bending over or who just want places to sit down in the garden, the raised bed you just built could be the foundation for a seating-height raised bed by following these steps. The height of the 2 × 8s (38 ×184 mm) is actually 7¼ inches (18.41 cm). So, two sets stacked up will be 14½ inches (36.83 cm) tall. The 2 × 6 (38 × 140 mm) cap is actually 1½ inches (3.81 cm) thick, so the finished height will be 16 inches (40.64 cm):
7¼" + 7¼" + 1½" = 16" (18.41 cm + 18.41 cm + 3.8 cm = 40.64 cm).

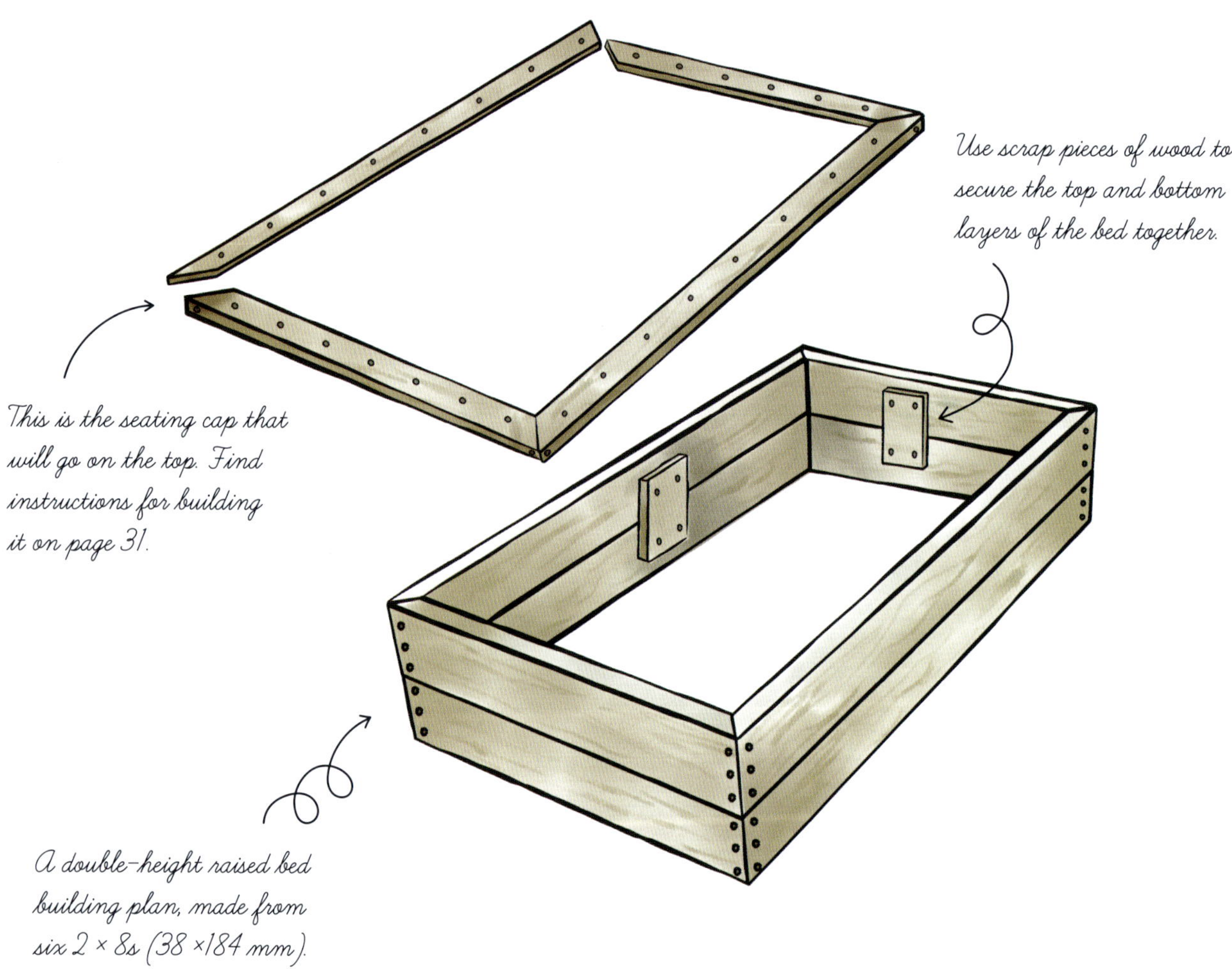

1. Simply build two identical raised beds of 2 × 8s (38 ×184 mm) as per the plan and instructions on pages 28–29, then stack one on top of the other.

2. Use scrap pieces of wood and screws to attach the top layer to the bottom layer. Use shorter screws so the pointy ends don't come all the way through. This will keep the layers aligned with each other.

3. Add 40 cubic feet (1.13 cubic m) of topsoil at this point. That way, your wheelbarrow won't bang up the seating cap.

Adding a Seating Cap

If you want to add a cap to the top of your filled bed for a nice place to sit, here's how:

1. Get three 10-foot (3 m) 2 × 6s (38 × 140 mm) to make a comfortable cap for sitting on.
2. Use a speed square to make a 45-degree diagonal line (not a bevel) to cut each end so it will resemble the corners of a picture frame. Cut each plank 4 inches (10.16 cm) longer than the piece of wood it will cover so it overlaps the walls of the bed by 2 inches (5 cm) on each side.
3. Lay the four cap pieces on a flat surface.
4. Drill one pilot hole slightly wider than the threads of the 3-inch (7.62 cm) screws, through the thickness each corner as shown in the illustration on the opposite page. There will be a total of eight holes.
5. With eight screws, fasten the corners together.
6. With a helper, lay the cap on top of the taller raised bed.
7. Position the cap so it sticks out 2 inches (5 cm) from the raised bed in every direction.
8. Use 3-inch (7.62 cm) screws to secure the cap to the top edge of the raised bed. With enough strength, you may be able to do this without drilling pilot holes. Check the position of the cap on the bed frequently as you work so it doesn't get out of alignment.
9. When the cap is installed, pour a cold drink and relax on your new seating-height raised bed!

Seating height raised beds—or any kind of beds—don't have to be in perfect rows. These beds are staggered and have an open space that allows party guests to sit in informal groups with their beverages and barbeque.

Wheel-Burro Wisdom

I recently enjoyed Andy Merrifield's book, *The Wisdom of Donkeys*. He hired a donkey to carry his baggage on a trek through the hills of France, and he really got into the head of his sturdy and reliable donkey. His journey inspired me to compile some tips on what I call *wheel-burro wisdom* because wheelbarrows are like a burro in that they carry our heavy loads. But they sure eat a lot less.

DELIVERY

I touched on it on page 25, but if you're having a load of soil, mulch, compost, or pathway gravel delivered for your tomato garden, put down a big, heavy-duty tarp first. The brown ones cost more than the blue ones, but they last longer and look less ghastly. It not only makes it clear to the delivery driver where you want the load dumped, but it also makes it easier for you to clean up at the end. You'll probably save 20 minutes by not having to sweep up the dregs of soil, mulch, or gravel.

A tarp is a target for the driver and a tool for making clean up much easier for you.

Also, I like to lay my tarp with the brown side up. The silver side has too much glare. Ditto for folding it up: Brown-side out looks slightly less garish than silver-side out.

Shoveling Material from the Tarp

Park the wheelbarrow on the tarp as you shovel so any spilled material will fall back on the tarp for easy cleanup.

Before loading the wheelbarrow, park it in the direction aimed at your goal. No sense having to turn a heavy wheelbarrow 180 degrees.

Remember, don't run your shovel or pitchfork too close to the tarp. You'll snag and tear it before getting even one good use out of it. Instead, leave a layer of material about 3 inches (7.62 cm) thick in place to protect the tarp from your tools. Once you've loaded all the material above that 3-inch (7.62 cm) layer, grab the edges of the tarp with both hands, lift it, and consolidate that thin layer of material into a bigger pile in the middle of the tarp. Then, repeat the process.

Loading a Wheel-Burro

Load it heavy over the wheel and light near the handles. With the bulk of the weight over the wheel, your load will feel lighter and be easier to steer when you're in motion.

I learned from a brick mason to position my tools, my supplies, and my work so I moved my body as short a distance as possible through the course of the day. He called the time spent unnecessarily walking back and forth "lost motion." Just his short demo and explanation changed my working style that day and every day since.

How can you avoid lost motion? Position the wheel-burro for loading on whichever side of the tarp so you are moving it the shortest possible distance. Over the course of an afternoon, this saves a lot of energy that you'll need later for mixing cocktails.

Humans are messy. Make cleanup faster by putting down a tarp before you load your wheel-burro. A tarp covering the bed of your truck before it's loaded will speed up the unloading too.

Traveling

Make wide, curving turns rather than trying to turn on a dime. You'll have more control, which means you're less likely to spill your load accidentally.

Focus on staying upright as your *primary* goal. Getting the payload delivered is your *secondary* goal. That way, your wheelbarrow will never tip over.

A wheel-burro tricked out as a mobile power source in Provence. The two charged-up batteries provided juice for power tools in the days before drills and saws came with their own batteries!

DUMPING

After dumping your load, sometimes it's easier to simply turn your body 180 degrees and pull the wheelbarrow behind you as if you were a donkey with a cart. When you park to refill it, the wheel-burro is already pointed in the right direction.

With some materials, like sticky topsoil, lifting and wagging the wheelbarrow side to side to fully empty it takes a lot of energy and sometimes you still have material that won't come out. In those cases, a wheelbarrow empties best if you lift it up by the handles and tip it all the way over until it's upside down. Then, walk around it to the far side, grab the handles, and roll it *sideways* until it's right-side up again—and magically pointed back toward the refill pile.

My wife, using a wheel-burro to clean out a donkey barn, in Provence on a WWOOF farm (World Wide Opportunities on Organic Farms).

MAINTAINING

If the tire pressure gets too low, a wheelbarrow becomes harder to push so keep a bicycle tire pump handy. Better yet, buy a solid tire that will never deflate.

Keep a wrench handy that's sized for the various nuts on the underside of the wheelbarrow. Even if they have lock washers or lock nuts that aren't *supposed* to come loose, they *will* come loose and fall off at the most inconvenient time. Ask me how I know this.

A PERSONALIZED EXPERIENCE

Don't be afraid to give your wheel-burro a personality: name it, paint it, "tag" it, tie ribbons on it, slap bumper stickers all over it, *something*. You two will be gardening together for a long time so it's wise to get on a friendly, first-name basis.

DIY Subirrigated Planters (Better Than They Sound)

I once built a couple of seating-height red cedar raised beds on the flat roof of an apartment building for a client. I'm sure they're doing well. But I wish I had known about my colleague Steven Biggs's DIY subirrigated planters (SIPs) for his tomatoes that thrive on the flat roof of his garage. This version is Steve's response to the smaller and more expensive store-bought version of subirrigated planters (a.k.a. "self-watering planters," although that's a misnomer—it's still up to the gardener to manage the watering).

ADVANTAGES

As a closed system, all the water and fertilizer remain available to the plants instead of running off as waste. Because the planter has its own reservoir of water beneath the planting area, you won't have to water as often. And the continuous moisture means your plants will grow at a steady pace, leading to higher yields. By growing in potting mix instead of soil, and getting water from below instead of above, the plants will also have fewer diseases. And since the soil surface will be drier, there will be fewer weeds. Being above ground in a dark container, the soil will warm more quickly in spring, boosting tomato plant growth.

LOCATING YOUR SIPS

If you don't have a sunny space in your yard for growing tomatoes, or if your soil is contaminated, these planters are a doable alternative. They also work on a flat roof, patio, deck, or driveway, if that's where you have room and if that's where the sun shines in your yard. Be sure to check the load-bearing capacity of your roof or deck, if that's where you plan to set these up. With all the water and soil they contain, SIPs are heavy.

THE BASIC IDEA

The concept is that the bottom of the container is filled with water (the reservoir), a length of 4-inch (10.16 cm) drain pipe, and a layer of weed fabric. The top of the container has potting mix, plant(s), and mulch. There's also a fill tube for adding water. An overflow hole in the bin keeps the potting soil from getting soggy. And one or more strands of polyester material—like acrylic yarn or nylon rope—wicks the water from the reservoir up into the high-quality potting soil. Here's your step-by-step pathway to growing tomatoes in a large container that also holds a lot of water.

MATERIALS AND CONSTRUCTION FOR BUILDING A SUBIRRIGATED PLANTER

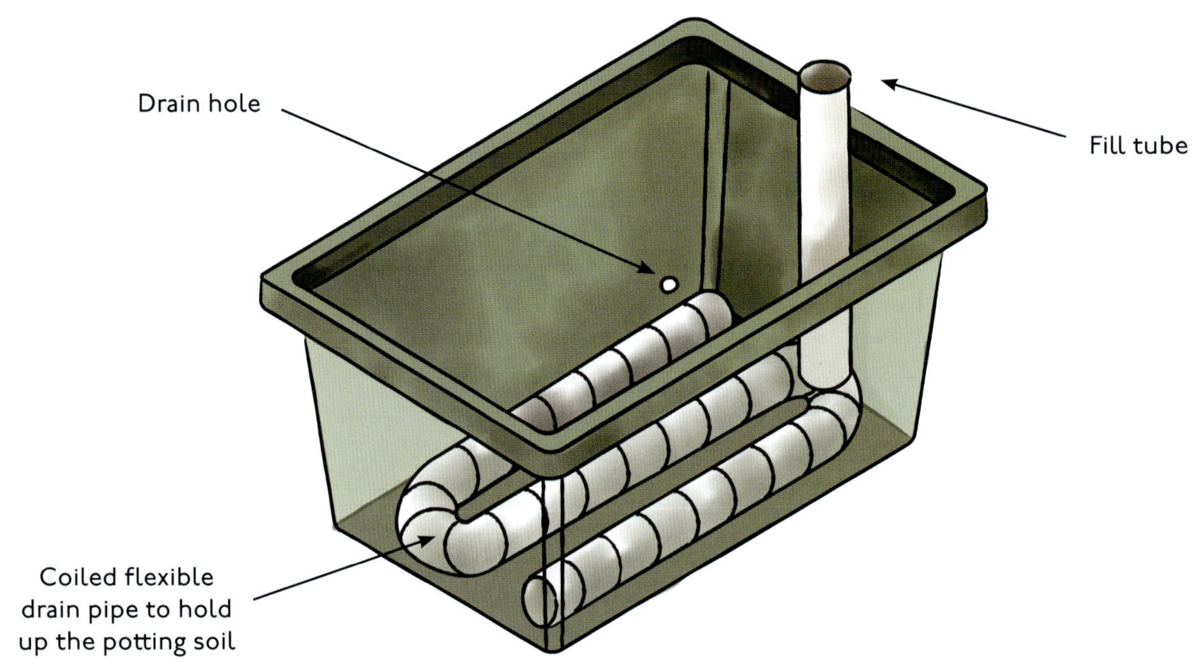

The Container

I like to use 27-gallon (102.2 L) storage totes from the big-box store. The dimensions are 12 inches (30.48 cm) deep, 12 inches (30.48 cm) across, and 24 inches (60.96 cm) long. They only cost about $10 each, and I have three of them positioned on my concrete driveway for growing dwarf tomato plants. The bottom third of the tote holds the water and the upper two-thirds holds the potting soil for the plant's roots.

Drain Pipes for the Reservoir

To create the reservoir and to hold the potting soil up above the water, use black, corrugated, perforated drain pipe that's 4 inches (10.16 cm) in diameter. There are various lengths and styles that sell for about $1 a foot. In a container this size, I use a knife or razor to cut three straight sections of pipe that are just under 24 inches (60.96 cm) long. I lay three of them side by side at the bottom of the tote. You could also buy a very flexible drain pipe that would fit the space without being cut (see illustration). Important: These sections of drain pipe aren't watertight; it's the tote's job to hold the water. The sections of drain pipe are open enough to let water in and also sturdy enough to hold up the potting soil. They are for structure, not for holding water.

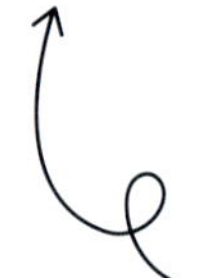

The 4-inch (10.16 cm) corrugated pipes don't convey water. They merely hold up the landscape fabric and potting mix, creating a space below, for the reservoir of water.

Drain Hole

With the drain pipes in place, put a ½-inch (1.25 cm) bit on a drill and make one drainage hole that's just above the drain pipes. The water reservoir won't be any higher than the drain hole. If the surface under the totes isn't level, position the end of the tote with a drain hole toward the higher ground. That way, when the tote is filled, some of the water will be in contact with the potting mix.

Tuck some landscape fabric around the refill pipe—like a scarf—to keep the potting mix from sliding down below the root zone.

Fill Tube

You will need a way to refill the reservoir underneath the mix after the potting soil is added. That's the fill tube's job. Black plastic flexible hose made for dishwashers is readily available, inexpensive, and easy to cut with a knife. Cut enough for one end to rest on the bottom of the tote and the other end to rise a couple inches above the top of the container's rim.

Wicking Fabric

I like to use a length of wicking fabric—nylon rope, acrylic yarn or burlap sack scrap. Coil about 1 foot (30.48 cm) of wick at the bottom of the tote, inside or outside of the drain pipes—doesn't matter. As the potting mix dries out, so will the wick. The dry, upper part of the wick will draw water from the wet, lower part of the wick. That's how the water in the reservoir gets drawn into the potting mix, where the tomato roots can quench their thirst. Scraps of cotton fabric also work as wicks, but they decompose over time. You can also forgo wicks entirely by positioning the landscape fabric and the potting mix in what my colleague Steven Biggs calls the "saggy diaper" arrangement. Allow a portion of the fabric and mix to sag down between the 4-inch (10.16 cm) drain pipes so a portion of the mix is always in contact with the water in the reservoir. That way, it's the mix itself that acts like a wick. A mix-wick, so to speak.

Landscape Fabric

Using scissors, cut a section of landscape fabric (a.k.a. "weed block," although it doesn't don't do that well) that will completely cover the drain pipes. Allow enough extra fabric on all sides to tuck it in snuggly where the pipes meet the sides of the tote. The fabric's job is to hold the potting mix above the reservoir of water. Make sure the fabric does not block the drain hole. Leave some space to tuck the weed fabric around the fill tube and your wicks.

Potting Mix

Don't skimp on potting mix. Your best choice is professional potting mix ("potting media" or "potting soil" all mean the same thing in practice). The better the mix, the longer it will hold water for the roots. If you can't lay your hands on professional-grade potting mix, see if you can find one with "pine bark fines," and mix it 9:1 with a little bit of builder's sand. If that's too much trouble, use the most expensive consumer potting mix you can find, using the cost as a proxy for the quality of material. You want a mix that can both hold plenty of moisture and plenty of oxygen. Whichever route you go, dumping your mix in a wheelbarrow and wetting it first will work a little better than putting it in the tote dry and then trying to wet it. You'll be able to mix it up in the wheelbarrow with a hoe or with your hands to get it thoroughly saturated (but not completely waterlogged); it will settle less as it dries out in the tote. Weave the wicks throughout the mix from bottom to top as you fill the tote with the mix. After filling the tote to the top with damp potting mix, settle it with your hands. Make sure to leave room for the fill tube. Or if you have access to enough, you can use decent compost as potting media.

To ensure that water is siphoned up into the potting mix, lay two foot-long (30.48 cm) strands of cloth—in this case free coffee bag burlap from a local roastery—in the reservoir and over the weed fabric.

Gently top up the container with potting mix—in this case compost—and gently settle it with your hands. You don't want to dislodge the landscape fabric. Leave room for the rootball of a tomato plant. In this case, we're adding a dwarf slicing tomato that probably won't need any kind of trellis to stay upright.

Any kind of mulch will keep the potting mix from drying out too fast and draining the reservoir. In this case, I've used a burlap coffee bean bag with a caffeine goddess to watch over things. Note the line from the right side to the tomato stalk. I cut there with scissors to allow installation of the sack without disturbing the tomato.

Mulch

Because you're using a potting mix, it won't contain any tomato diseases. But if you bought your transplants from a nursery, they may bring tomato diseases from that grower's greenhouse. So, even in a container, make sure you have a layer of mulch to keep rain water from splashing tomato diseases from the soil onto the lower leaves. You'll also want mulch because squirrels will see the loose soil as a possible place for finding or storing acorns. Plus the traditional virtues of mulch also apply: It keeps the soil from drying out too fast and keeps weeds from colonizing that container. Black plastic mulch is a great option for SIPs. With scissors, cut the mulch wide enough so you can tuck in the excess all along the edges of the tote. Leave room for the fill tube. Cut a 3 × 3-inch (7.62 × 7.62 cm) square in the plastic mulch to give you room to plant through the plastic and to keep the plastic from touching the plant stem in summer. After planting, add a big handful of organic mulch to cover the opening's exposed soil. Any other mulch you're using—leaves, chaff, pine bark fines, burlap coffee sacks, etc.—will work well on SIPs too, but you won't get the advantage of the early soil warming that black plastic provides.

WATERING YOUR SIP

You SIP is now complete and planted. If you're using a watering wand to refill the totes, unscrew the part that's called the "rose" and the end of the wand will fit in the dishwasher fill tube so you're not spilling water on the leaves of your tomato plants. Top up each tote until excess water shoots out the drain hole. How often will you have to refill the tote? That depends on air temperature, the size of your tomato plant(s), and the absorbency of the potting soil and wick. You may have to refill it once a week or so depending on these variables. You can also automate the watering by using what's called a "spaghetti" tube attached to a drip line to put water in the fill tube. With the drip line attached to an automatic timer on the spigot, you can set a schedule for refilling.

NEXT SEASON

Here are three end-of-season steps for getting ready for next year's crop:

1. Drain the water from the SIP with an inexpensive plastic siphon from a big-box store or a marine supply shop. This will make the SIP lighter and easier to move for the off-season.

2. Pull back the mulch and water the potting mix until the reservoir fills again to drain the mix of any remaining fertilizer salts. Use the siphon again to remove this water. While not toxic, over time, fertilizer salts build up, so better to stay ahead of that.

3. Since potting mix is made of organic materials—pine bark fines, peat moss, etc.—that will slowly break down into carbon dioxide and water and the quantity of the mix will decline. Top up the potting mix in spring before next season's planting.

Let *Your* Tomato Pots Hold More *Roots*

If you're growing tomatoes in regular pots, forget adding a base layer of gravel or broken crockery to "help drainage." With modern potting soils, it's unnecessary; think of the potted plants you bring home from the nursery: There's no "drainage layer" in them. A "drainage layer" only serves to reduce the volume of potting soil used in the pot, which then reduces the quantity of roots, which then increases your frequency of watering. The practice of adding "drainage" probably stems from when gardeners had only actual, heavy garden soil to put in their pots. Some pieces of broken crockery *would* help in that case. But modern "soil-less growing media" is designed to drain very well. So, save yourself the effort and give your plants more room to root around. And, maybe, use that "drainage" gravel as mulch on top of the potting soil instead.

This photo shows how adding "drainage material" at the bottom of a pot, merely reduces the amount of potting mix for roots to grow in. A pot set up like the one on the left, will grow more roots and a bigger, healthier plant.

Growing Tomatoes in Straw Bales

Is your only sunny spot a driveway or patio? Is your soil too problematic for a conventional ground-based garden due to pollutants, poor soil, or rampant weeds? Do you prefer gardening from a standing position? Do you want to start gardening with a disease-free tomato garden? Straw bale gardening does require more money, time, and water than gardening on the ground, and your first time doing it will involve a small learning curve. But, if you are comfortable with all that, straw bale gardening could be for you. Done right, you will be able to grow a lot of tomatoes.

STRAW BALE PREP

Only use straw bales for this kind of planting. Hay bales and pine needle bales will give you poor results. You can buy straw bales from farmers, farm supply stores, and some garden centers and big-box home supply stores. Most of these places will also deliver.

Popping your tomato transplants into brand-new straw bales would also be a mistake. Raw straw holds very little in the way of moisture and fertility. It would almost be as bad to leave your plants on a table in the sun. After placing the bales in your growing site, you'll need to allocate two weeks before planting day to stimulate some root-friendly decomposition in the heart of each bale. It's a process called "bale cooking."

With her plants well established, Zoe removes the thermometer from her long row of straw bales. The determinate tomatoes are held up by a cat's cradle trellis with stakes driven through the bales and into the ground.

With your bales in place, add water every day for two weeks—enough that you can see the excess draining away from the bottom of each bale.

Every *other* day, add fertilizer to each bale. If you're adding dry granular fertilizer—whether organic or chemical—add water afterward to help the fertilizer drain down through the straw and into the heart of the bale. If you're using liquid fertilizer, and make the mistake of adding it first, the water will just dilute it and push it out the bottom of the bale. You would have the same problem when fertilizing and watering any containers. So, add liquid fertilizer *after* watering, or add dry fertilizer *before* watering.

With both kinds of fertilizer, choose something with a high nitrogen level: the first number on the bag. Chemical fertilizers for lawns are a good option (but forgo any that also include herbicides). They will have 15 to 46 percent nitrogen. Add about 1 cup (weight varies) to each bale, every other day. With organic fertilizers, the nitrogen will range from 5 to 12 percent. Add about 3 cups (weight varies) per bale, every other day.

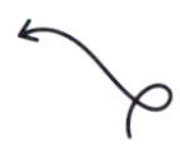

If you can't confirm that your local straw bales haven't been treated with a long-lasting herbicide, buy one bale a month early and sow a cold weather plant like snap peas in the bale. If they sprout and grow for a couple weeks, then your bales are herbicide free.

As you work through this process, use a meat thermometer or any kind of digital thermometer with a probe at least 6 inches (15.24 cm) long to check the temperature of the bale each day. At first, the temps will resemble the air temperature, perhaps around 80°F (26.6°C). Once the nitrogen starts stimulating decomposition, the temperature will rise to as high as 120°F (48.8°C) from the heat given off by active microbes that are reducing the raw straw into compost. By the end of week two, the temperature should be dropping back to roughly match the air temperature. You're now ready to plant.

PLANTING YOUR BALES

I placed nine bales to form a patio-like enclosure (more details following) on my driveway and added a 1- to 2-inch (2.54 to 5 cm) layer of compost to cover the tops of the bales. Some roots will grow horizontally through the compost and some will grow down through the bale.

Each bale has room for up to two plants. I placed one dwarf tomato plant in each bale. Their thick stems and short stature mean they need little to no support to stay upright. And if they do flop over, they are not near any disease-ridden soil. At the other end of each bale, I put in sweet potato plants, 'Crimson Sweet' watermelons, cucumbers, or a couple of grape tomato plants. The watermelon, cucumbers, and sweet potatoes are allowed to drape over the tops and edges of the bales to make them more attractive. The watermelons and cukes will be harvested from the driveway over the summer. The sweet potatoes will be harvested in late summer as the bales fall apart. The grape tomatoes will sprawl on top of and down the sides of the bales, for convenient snacking.

After adding a layer of compost on top of each bale, use a trowel to pry open a wide slit in the bale to insert more compost, organic fertilizer, and the root balls of your transplants. Tuck the tomato plants in as deep as you can get them, while leaving the top leaves visible. Hand water all your plants when you're done planting.

To keep the leaves as dry as possible, install a soaker hose alongside each plant without it touching the stems. Where the soaker hose exits the straw bale garden, attach a normal garden hose and run that to the spigot. Between the spigot and the garden hose, install a water timer. If your budget allows, use a battery-powered timer you can set on a schedule.

Unlike an in-ground garden, it's impossible to overwater a straw bale, so err on the side of too much water. Try running the timer for 20 minutes each day. You want to see water running out the bottoms of the bales. And you don't want to see wilting plants—unless it's the afternoon of a very hot day. Wait to see whether the plants recover in the morning before adding more water.

After placing your soaker hoses, cover them and the top of each bale with a 1- to 2-inch (2.54 to 5 cm) layer of mulch. I use triple-shredded hardwood bark mulch. It has a fibrous texture that knits together and doesn't wash away in the rain. Cover the entire top surface of each bale to create an attractive, finished look and to reduce water loss from evaporation by the sun.

STAKING YOUR TOMATO PLANTS?

Bales are not as dense as soil, so they don't support stakes or trellises very well. If your bales are on the ground, you could drive stakes through the bales and into the soil to stabilize them. My bales are on a concrete driveway, so that's not an option. That's why I chose dwarf tomatoes that can stand up unsupported and grape tomatoes that can flow over the bales and the driveway without catching soil-borne diseases.

I use a sharpshooter shovel to plant tomatoes in the bales. The blades are long and narrow. Insert the blade into the bale, tilt it back, slide the transplant into place and hold it while you remove the shovel. Press the bale and compost snug around the rootball and stalk of your tomato plant.

STRAW BALE GARDEN MAINTENANCE

Keeping a straw bale garden going over the growing season is pretty straightforward. You'll have few to no weeds because there's no soil. You will have some wheat grains from the bales that will germinate, but they add charm without competing with your tomatoes as they will die out in the heat of summer.

Diseases will be few to none, as those are soil-borne and your plants aren't growing in soil. So, leaving them unstaked and sprawling is an option.

After setting transplants in the straw bales, add a thick layer of compost. Then, place a soaker hose near the tomato roots, and cover the entire tops of the bales and the soaker hoses with a two-inch (5 cm) layer of mulch that won't wash away in the rain, such as this shredded hardwood bark mulch.

If plants are wilting in the morning, the roots are dry and you need to bump up your watering volume or length of watering time.

Any bug problems will be the same as if the plants were growing in the ground, so "WALA WALA"—which means walk around, look around—two times a week, examining your garden for signs of any bugs or vermin (see chapter 6 for more on managing tomato pests).

If the leaves get pale green to yellow, there's not enough nitrogen fertilizer or what you added at the beginning has been leached out by the rain and irrigation. No harm in making up the difference by using a liquid fertilizer applied to the base of each plant every week.

DECOMMISSIONING YOUR STRAW BALE GARDEN

By the end of the growing season, the straw bales will be a grayish-brown color and looking pretty defeated and deflated. After you've pulled your plants, you can pitchfork the broken-down bales into a wheelbarrow and then onto your fall raised garden beds for mulch or into your compost bins. Remove their plastic twine and toss that in the trash as you go. After you've cleared away your old bales, you could set up new ones for a cool-season garden. Or make notes about any improvements or changes for next year's tomato crop.

A wide range of Zoe's cherry and grape tomatoes grown in her straw bale garden.

LET'S
do a
PROJECT

A Tomato Snack Patio

MANY GARDENERS SET UP THEIR STRAW BALES IN A LINE LIKE SO MANY TRAIN CARS. AND THAT WORKS FINE. BUT THAT'S NOT THE ONLY OPTION. BECAUSE OF THEIR HEIGHT, THEY CAN BE ARRANGED IN A RECTANGLE—OR EVEN A CIRCLE!—TO CREATE AN ENCLOSED SPACE THAT FEELS LIKE AN OUTDOOR ROOM.

Here are the steps to create your own straw bale tomato snack patio:

ABC: ARRANGE BALES COHERENTLY

As shown in the photo on the next page, I arranged nine bales of wheat straw to form a rectangle with a walk-through opening. Nine or ten bales is the minimum number needed to create an enclosure for two people to sit comfortably.

Before deciding how many bales to buy, I recommend placing at least a couple outdoor chairs in your space and using them to define the size, the views, and orientation of your snack patio.

With your chairs in place, use some bricks or rocks to mark the corners and an opening that would otherwise be filled by one bale. My opening could be filled by two bales, but I left it that wide so I could further define the entrance by using a couple of salvaged terra cotta chimney flue sections to hold flowering plants.

Using a tape measure and—ideally—a graph paper notebook, draw up the dimensions and work out the placement of your bales. Allow 2½ to 3 feet (76.2 to 91.44 cm) for the length of each bale. Make adjustments on the ground and on your graph paper until you have a floor plan you're happy with. Then, buy some bales. It's not a bad idea to buy one more bale than your plan calls for, just to account for any variation in size or last-minute alterations to the plan. If you don't need the extra bale, it can serve as mulch in the vegetable garden or as bedding for someone's backyard chickens.

DEFINE THE ENTRY

Look at any front door on a house or any gate in a yard and notice how they often have details that draw attention to them: intricate trim and heavy hardware for a door and perhaps an arch and a fancy handle for a gate. These details are *meant* to draw your eye to these entrances. If nothing draws your eye there—if the door looks like a wall and the gate looks like more fence—how would you know where to get in? Plus, those details just look cool. For all those reasons, the opening to your snack patio also benefits from this kind of attention.

You could place a couple of large potted plants, stacks of bricks to support smaller potted plants, or salvaged chimney flues to hold potted plants as I've done at the opening. Between these features, be sure to leave at least a 3-foot (91.44 cm) wide gap—about the width of most doors and gates—so the entryway doesn't feel crowded.

To make my straw bale tomato garden look more appealing and to create a new garden space, I arranged them in a rectangle with a pedestrian opening. Chairs, side tables, potted plants, indoor/outdoor carpet, and a door mat define the space as a "garden room." A neighbor's donation of a roll of vintage fence wire inspired me to line the inside and outside of the bales with a supporting layer to keep them from sagging. And it looks pretty cool, too. This photo was taken before the straw bale garden was planted.

TAKE A SEAT

I pulled a couple of outdoor chairs from our front porch into this space to make it comfortable for coffees, cocktails, or cherry-tomato snacking with company. Use chairs that can stand up to the weather, are comfortable, and look good to you. Use just one chair or as many as you can fit comfortably.

Echo and Dyson—surrounded by dwarf tomatoes, grape tomatoes and sweet potatoes—catching up over coffee in the straw bale tomato snack shack in my driveway.

ENHANCE THE SNACK SHACK

When a neighbor walking her dog asked if I wanted a roll of vintage garden fencing she'd never gotten around to using, I headed right over with my hand truck to bring it home.

I wanted to make the snack patio look even more gardenesque. I used my bolt cutters to trim the length and height of the fencing so it would fit around the outside of the bales. I attached the fencing to the horizontal twine around each bale, using zip ties. Twine or tie wire will work as well if that's what you have handy. The fencing isn't structurally absolutely necessary. But, by the end of the season, the bales will be sagging a bit. So, vintage fencing like this, or any kind of wire fencing, even chicken wire or wooden lattice, will keep your bales from looking quite so tired in the latter half of the season.

Since our snack patio is on a driveway, I put down an inexpensive 6 × 8-foot (1.82 × 2.43 m) outdoor rug to better define the space as a welcoming hangout. If your snack patio is on soil, such a rug would also function as a mulch to keep down weeds. Alternatively, you could use a layer of cardboard and any kind of organic mulch to define the floor, suppress weeds, and keep the ground from getting muddy. And, it doesn't hurt to place a spare welcome mat at the entrance for an inviting feel.

Once all those elements are in place, you can start having family, friends, and neighbors hang out with you in your straw bale tomato snack patio. You don't even need to wait for the tomatoes to get planted, much less to be in fruit. You just need some decent weather, the right company, and the right beverage.

LET'S *do a* DEEPER DIVE

Raised Bed Revolution by Tara Nolan covers different styles of raised beds.

Straw Bale Gardens Complete by Joel Karsten introduced the gardening world to the notion of growing in straw bales.

Foodgardenlife.com is the website for my gardening colleague Steven Biggs, who came up with the DIY version of the self-watering container outlined in this chapter (see p. 35).

CHAPTER 3

Organic Ingredients: Fertility, Water, Oxygen, and Sunshine

AT SOME POINT IN ANCIENT HISTORY, THERE WERE NO TOMATO PLANTS (!!!), BUT EVOLUTION LIKES TO TRY NEW THINGS AND ROLL THE DICE TO SEE WHICH NEW CREATURES WIN OUT.

Our planet has existed for over four billion years. For roughly the first three billion-ish years, it was only a giant rock with thunderstorms that poured down into lifeless seas. About a billion-ish years ago, microorganisms—that were the ancestors of plants—mysteriously started growing in wet floodplains. What they had in common with all of our modern plants is that their food supply was sunlight. And they needed the same array of molecules for cellular building blocks, including:

- CO_2 (carbon dioxide) from the air, absorbed by the leaves
- O_2 (oxygen) from the air spaces in the soil, absorbed by the roots
- A smattering of minerals, like nitrogen, phosphorus, potassium, and others, in the soil, absorbed by the roots
- H_2O (water) from spaces in the soil, absorbed by the roots

The modern plants that evolved from these microbes use the real-life magic of photosynthesis to combine these ingredients to make sugar with which to build their cells and power their growth.

Lucky for us, as a part of this process, all leaves "poop out" a marvelous waste product called oxygen. This was bad news for some of the earliest microbes to whom oxygen was poisonous, however. Many of these anaerobic microbes died out—the first great extinction—and the rest hid in crannies in the soil and marshes without oxygen. But this was good news for the animals that were soon to evolve—they required oxygen. Granted, some readers may be disturbed to learn that the oxygen we breathe is a waste product. Sorry. But what goes around comes around.

Soil Fertility

A brilliant organic gardening concept is to "feed the soil, not the plant." This means that if you've provided the garden soil with:

- Plenty of organic matter
- A wide array of nutrients
- A layer of mulch to protect the soil from compaction by rain and dehydration by sun . . .

. . . your tomatoes will be able to grow plenty of leaves, stems, roots, flowers, and fruit while developing as good an immune system as possible for fending off bugs and diseases.

Where soil test boxes go to die. After soil scientists have tested your carefully collected sample, it all just goes into a dumpster for composting.

SOIL TESTS YOU DON'T HAVE TO STUDY FOR

Agricultural extension service offices affiliated with universities here in the United States offer soil tests for free or little cost for gardeners and farmers. These are worth taking advantage of, especially for novice gardeners and even for experienced gardeners who've never done one. You may be missing some critical information about your soil. These tests come with a sample collection container or bag and a sheet of simple instructions. Get one test for each separate tomato bed. You can return the filled collection container to the extension office and expect to receive your results in a matter of weeks, depending on how busy the lab is.

A couple of caveats:

The test results will give you a recommendation on how much nitrogen your garden needs, but it is just an estimate. The test for nitrogen is expensive and you have to ask for that separately. The directions don't always make this issue clear. Also, you'll be tempted to dig one scoop of soil from one spot to fill the container. But if a bunch of fertilizer or lime was spilled on that spot in the past, the results for that bed will be skewed. Instead, take small samples from around the bed to fill the collection container. When you get the results, recognize that there is a margin of error. None of the numbers are precise; they're just close enough to give you a sense of what will need to be added to amend your soil and roughly how much.

FERTILIZERS FOR PLANTS EQUALS PLANT FOOD, RIGHT?

Nope. Fertilizers—whether organic or synthetic chemical salts—are more comparable to vitamins for humans.

Only sunlight is food for plants. Consider this if you're thinking of growing tomatoes in a spot that gets less than eight hours of sunlight.

When I become the world's philosopher-king, one of the first things I'll do is declare it illegal for fertilizer companies to put the term "plant food" on the bag.

SUNLIGHT

The *only* food for plants is light, which they use to magically stitch together water from the ground and carbon dioxide from the air to make the equivalent of table sugar. That's what the word "photosynthesis" means. Just try to imagine that happening in every living leaf as you look around outdoors. Every leaf is magically making sugar: Twelve molecules of CO_2 + eleven molecules of H_2o = one molecule of sugar $C_{12}H_{22}O_{11}$ that goes to the living parts of the plant + twelve molecules of O_2 that it expels into the atmosphere.

Once you see that, then the question is, "How much sunlight are my plants hungry for?" If you've ever tried growing vegetables or herbs indoors—on a windowsill, perhaps—you've seen what plants look like when they are on a starvation diet. They are really unhappy. They will produce some leaves, but not a lot and not enough to support much fruit. It helps to think of fruits as an expensive investment by any kind of plant.

As for nutrients from fertilizer, compost, manures, wood ash, compost, and lime, I think of those elements as being similar to vitamins for people. We can't live off vitamins alone. But when we get them consistently in the right proportions, we can live *better*. Ditto regarding nutrients for plants. Let's go over the virtues of various nutrients.

The sunniest spot on our property is the front yard. So to make sure our vegetables don't go hungry, we built our diamond-shaped raised beds there. The right-hand bed features indeterminate tomatoes in four tomato towers. The grassy spot nearby is what my wife Chris calls "The Lawnlet." On the far left is our herb garden.

Mediterranean regions evolved using fires—caused by lightning—to return nutrients to the soil and raise the pH. The oils in Mediterranean herbs both deter insects and make the leaves flammable, as seen in this lavender plant in my garden. Fortunately, you don't have to burn your garden to raise the pH. Some lime will do that.

LIME

Lime is pulverized Dolomitic limestone from a quarry. It contains the two macronutrients calcium and magnesium. Before the lime was compressed into stone, it lay underwater as a thick layer of shells of ancient ocean creatures similar to oysters, clams, and mussels. Tectonic plates compressed the shells into limestone and then shoved some of those underwater stockpiles above sea level. There, lucky landowners mined their land for something valuable to every farmer and gardener.

Added to the garden, lime raises the pH of the soil to a level closer to the pH of the west coast region of South America—Peru and Chile—where tomatoes were designed by evolution. Growers in Mexico, Spain, Italy, and other Mediterranean regions do not have to add lime to grow robust tomatoes. Those low-rainfall regions have a similar pH to the region tomatoes evolved in. The reason pH is important is that a given level of acidity is like a key that can "unlock," so to speak, certain soil nutrients for plants. In moderate-rainfall areas such as eastern Asia and eastern North America, and in higher-rainfall areas such as the US Pacific Northwest and northwestern Europe, the extra rain makes the pH lower and tomatoes are less able to access the nutrients in the soil that they need. So, adding lime artificially creates the Mediterranean-ish soil conditions tomatoes need to thrive. The free or low-cost soil test from your local agricultural extension office I mentioned previously will also help you determine the optimal amount of lime to add to your soil to make soil nutrients available to your tomatoes.

ORGANIC FERTILIZER VERSUS SYNTHETIC CHEMICAL FERTILIZER

Synthetic chemical fertilizers are a type of salt, but not one you'd want to eat. In this case, "salt" means it will quickly dissolve in water, like table salt. It also means that if this fertilizer touches any part of the plant, its saltiness will draw moisture out of the roots, stems, or leaves and leave them looking burnt. That's why the text on the bag is so adamant that you not let this chemical salt touch the plant. Be forewarned.

Because chemical fertilizer dissolves so quickly, plant roots can absorb water from soil that contains small amounts of this salt and put its nutrients to work right away. The quickest response is from the nitrogen, which is the first of three numbers on any bag of any kind of fertilizer. That's why gardeners see such quick results—greener, bigger leaves—when using synthetic chemical fertilizer. It gives quick, positive feedback. But that feedback can be misleading. That quick feedback from nitrogen for plants is somewhat similar to the quick feedback you'd get from feeding sugar to a child: They may be racing around on a sugar high, and they may be putting on weight quickly—probably too quickly—but, at some point, they're going to crash and their health will suffer.

The comparable problem for your plants from too much nitrogen is that the rich color is like a sign flagging down insects and diseases that says: "This plant has grown so fast that its immune system has been diluted and is less of a hazard for you, so come on over and chow down."Plus, if it rains a lot, the synthetic chemical fertilizer salts will be carried down into the soil beyond the reach of the roots. That's why the directions on the bag often recommend "side dressing" with more fertilizer one or more times in the growing season. As if you don't already have enough to do.

Organic fertilizers, on the other hand, are like a balanced meal that strengthens the tomato plant's immune system and doesn't leave it hungry for snacks before the season is over. The numbers on the bag of organic fertilizer may be lower, but don't let that concern you. It's an issue of the high *quality* of organic fertilizer versus the too-high *quantity* of synthetic fertilizer. Synthetic chemical fertilizers will contain only the three macronutrients—nitrogen, phosphorus, and potassium—listed on the bag. It's not unlike a person who eats only a few of the same items at every meal. Organic fertilizers, because they are sourced from a range of ingredients, contain those three macronutrients, plus dozens of micronutrients. Such a vast variety boosts the immune systems of tomato plants and enhances the flavor of tomatoes.

Let's meet these plant nutrients and discuss how they help your tomato plants grow.

NITROGEN

The first number on any bag of fertilizer represents its percentage of nitrogen. All parts of the plant need nitrogen, but the leaves are very dependent on it. If your plant's leaves have a yellowish pallor, the likely cause is lack of nitrogen. The remedy is to add fertilizer to the soil. Rather than spreading it on the surface and waiting for it to be slowly soaked into the root zone by rain or irrigation, I recommend raking a bit of mulch back, digging a shallow hole, and dropping a handful of organic fertilizer in the hole where the roots can quickly reach it. Put the soil back in place and replace the mulch. Dump some water on that spot so the roots will be attracted to it and start sucking up nutrients.

Sources of nitrogen in organic fertilizer are often waste products from various food-related industries. If these "waste" products didn't go into organic fertilizer, they likely would end up in a landfill creating methane gas that speeds up climate disruption even faster than carbon dioxide does. So, using organic fertilizer is also a small way to fight climate disruption. Sources might be cottonseed meal, alfalfa meal, fish and crab meal from seafood processing plants, and bloodmeal from slaughterhouses.

Nitrogen in synthetic chemical fertilizers, on the other hand, is extracted from the air in an expensive process that requires lots of fossil fuels that produce the carbon dioxide that boosts climate disruption.

In the off-season, you can grow a crop of legumes, such as crimson clover (shown), peas, or beans. Legumes collaborate with microbes to draw nitrogen gas from the air and build their cells with it. Once they mature, legumes can be tilled into the ground to release that nitrogen for your tomato crop.

PHOSPHORUS

The second number on the bag is the percentage of the nutrient phosphorus. All parts of the plant love them some phosphorus, but the flowers and fruits are most dependent on it. If your tomato plant is producing few flowers, the three most likely causes are 1) the air is too hot, 2) a lack of plant food (a.k.a. sunlight) because flowers are more costly to produce than leaves, or 3) a lack of phosphorus. If you can't do anything about the heat or sunlight situation, pop a handful of rock phosphate, bonemeal, or an organic fertilizer in a shallow hole in the ground where the roots can reach it.

Bonemeal is a waste product from slaughterhouses that would go to the dump and boost climate disruption if organic gardeners didn't use it. Rock phosphate is mined from quarries where ancient rivers and ocean currents deposited animals and fish for millions of years. Their bones and teeth compose much of the phosphorus deposits that were lifted above sea level by tectonic plates. These phosphorus quarries are great sites for finding shark's teeth and other fossils too. But ask permission first, before exploring.

Finally, in the cool weather of spring, your plants may have some trouble moving phosphorus through their leaves. They'll develop a purplish cast. Not to worry. It's a temporary condition that resolves as the weather warms.

POTASSIUM

The third number listed on a bag of fertilizer is potassium. All parts of a plant make use of potassium, but it's especially important for root growth. Potassium in organic and synthetic chemical fertilizers is mostly quarried from places where ancient seas dried up, leaving deposits of potassium.

Sometimes, potassium is called "potash," but that's a form of potassium not available on the market anymore. In the 1800s, people leached potassium from wood ash in big pots. Hence the descriptive term "pot ash."

Fertilizer *from* Firewood

They say firewood warms you four times: 1) when you cut it, 2) when you stack it, 3) when you carry it, and 4) when you burn it. In my experience, it can warm you three more times: 5) when you scoop out the ashes, 6) when you spread the ashes in your garden, and 7) when you get a warm glow from eating the vegetables that grew so well thanks to the ashes improving the soil.

If you live in high rainfall areas where the soil is acidic (here in the United States, that's mostly the east, the Midwest, and the Pacific Northwest), you probably spend money spreading lime for your vegetables, fruit trees, herbs, asparagus, and lawn, but you could save money and keep wood ashes from going to waste by using them as a substitute for lime and potassium fertilizer.

I once lived in a house with half a dozen crepe myrtles shading the backyard. I wanted to grow vegetables and fruit trees, and since the crepes bloomed with an offensive bubblegum-pink color, I cut down two every winter and burned them in my woodstove the following winter until they were gone. I added the ashes to the vegetable garden.

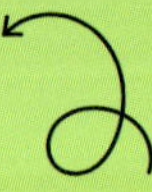

A wood stove, fireplace or firepit will produce an organic fertilizer, otherwise known as wood ash. Only handle it once it's cooled down.

Those wood ashes helped improve my soil in three ways:

1. They added potassium and other micronutrients to it. In fact, one of our nation's first exports to England was wood ash, sold as a potassium fertilizer. If wood ash came in a bag it's N-P-K (nitrogen-phosphorous-potassium) ratio might be as much as 0-1-10.

2. In addition to nutrients, wood ashes are alkaline (that's why they're also used to make lye and soap) so their presence can make nutrients in an acidic soil more available to plants and boost microbe populations. Wood ash has about half the calcium carbonate as lime: That's what raises the soil pH. So, if your soil test says to use 100 pounds (45.35 kg) of lime, you could use about 200 pounds (90.71 kg) of wood ash for roughly the same effect. A 5-gallon (18.92 L) bucket holds about 15 to 20 pounds (6.8 to 9 kg) of (cool) wood ashes. It's dusty and caustic stuff so wear a bandana or a dust mask and gloves while spreading it by hand or with a lime spreader.

3. I throw any leftover chunks of charcoal from the woodstove into my compost where they provide reserves of oxygen for composting microbes.

A couple of caveats: Don't apply wood ashes to acid-loving plants like blueberries and potatoes; don't apply near seedlings as it's too salty; don't let it get wet or drop it in clumps as it can be as caustic as bleach in high concentrations.

What should you do with your ashes if you already have a neutral or alkaline soil that would be made worse by applying wood ash? Maybe it's time to Google a recipe for homemade soap!

Chunks of charcoal are best added to the compost bin. They boost the quantity of oxygen and microbes in the compost and the plant beds. Again, only handle after they've fully cooled down.

CALCIUM AND MAGNESIUM

Calcium and magnesium aren't in most bags of synthetic chemical fertilizer, but they are the main components of Dolomitic lime.

If the plant's leaves are yellow between the veins, that could be a sign of magnesium deficiency. You may need to apply Dolomitic lime to the soil.

Early tomato fruits that have a black spot on their bottom—called blossom end rot (see p. 110)—have traditionally been considered a result of a lack of calcium, but it's better understood now as being caused by a lack of the consistent moisture needed for the plant to move calcium all the way to the fruit. To solve blossom end rot, pump up your irrigation. Even if you don't add more water or calcium this season, the problem will resolve itself as the plant gets bigger and stronger and its roots reach deeper.

MICRONUTRIENTS

In addition to the five macronutrients just discussed, tomato plants need these five micronutrients:

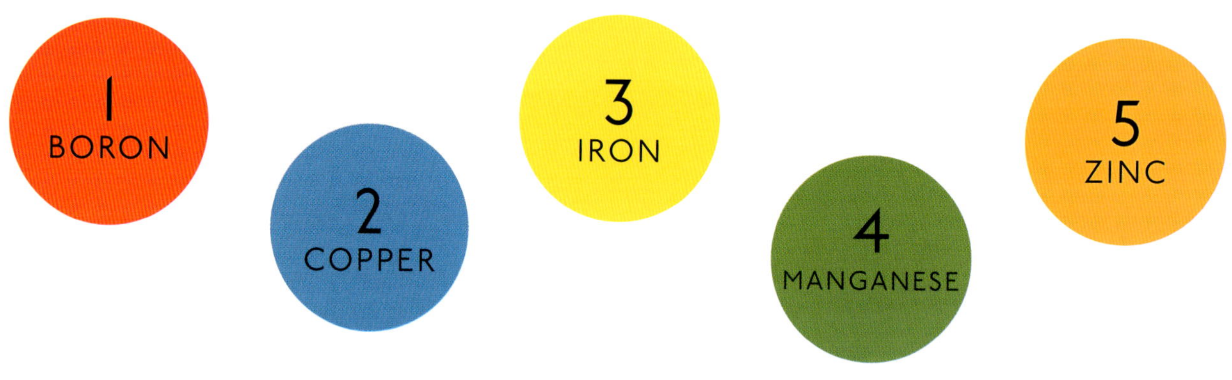

They all play important functions and are rarely found in chemical fertilizers. But if you're applying organic fertilizer, compost, or manure, your micronutrient bases will also be covered.

MANURE FOR THE GARDEN

Lots of our domesticated animals—cows, especially—are raised in covered stalls. This means the animals don't spread their manure across pastures, as in the good old days when organic farming was the only option. Instead, the manure piles up under the covered stalls. Yes, yucky. But some farms manage to get that manure composted, dried, and bagged up to sell to gardeners. The bag will display the three same N-P-K numbers that fertilizer bags display, but the quantity will be so low—usually 0.05 percent-0.05 percent-0.05 percent—that it's not enough to help your tomatoes grow. So, why bother? Well, because your tomato plants aren't the only living things in the bed. Manure is a food source for earthworms and microbes in the soil, and in return, those organisms will:

- Loosen the ground to allow more oxygen in the soil
- Make micro- and macronutrients more accessible by pooping out these minerals in a form that roots can more easily absorb
- Sometimes, outcompete bad microbes that would attack your tomato roots and leaves

A bonus: The manure itself will absorb and hold water from rain or irrigation and keep it available for thirsty roots.

And a caveat: Just because farm animal manure is good for the garden doesn't mean that manure from house pets can be added as well. Cats and dogs are carnivores and their poop . . . okay . . . manure . . . can hold diseases that humans are susceptible to. Plus the stink is hard to get past. Do not add it to your garden or compost.

COMPOST

When you dump compost in your garden beds, the results are very similar to adding manure. In fact, the process of composting is vaguely similar to what happens in the gut of a cow: Plant matter is broken down by microbes into a crumbly, brown, water-, microbe-, and nutrient-holding garden ingredient. When you add compost to your garden, you're getting the same benefits as just described for manure (more on composting in chapter 10).

ORGANIC MULCH

After planting tomatoes and laying down a soaker hose, a good gardener will apply a layer of organic mulch. Not only does this mulch keep the soil from drying out and keep bad microbes from bouncing onto lower plant leaves with raindrops, but it decomposes over the course of the season and plays a role similar to compost and manure: feeding the earthworms and microbes that loosen the soil, making nutrients more available and combatting the bad microbes.

Water

Humans and other mammals have a bloodstream that distributes nutrients throughout the body. That stream of blood also helps keep us cool. Our skin makes sweat from our blood. Plants use water to do the things our body uses blood for. Water distributes nutrients throughout the plant, and it helps cool the plants a bit when it's expelled through the leaves. When liquid water in the plant is turned into vapor as it exits the leaves, the vapor absorbs heat from the plant. Well, a smidgen. But every little bit counts. Water does one other valuable thing for plants: It plays the role of our skeleton in holding the whole contraption upright. Plant cells full of moisture act like a water balloon. They can hold things upright contrary to gravity. So, when a plant is low on water, it droops.

A caveat: A plant wilting in the afternoon sun isn't necessarily low on water. Even humans wilt a bit in the afternoon when it's hot enough. With enough sun blasting on it, a tomato will let its leaves droop for two reasons: 1) A drooping leaf is less exposed to direct sun, so it's not quite as hot, and 2) a hot leaf can be transpiring water at a faster rate than the roots and stem can provide a re-supply. Before bumping up the irrigation, wait to see if the leaves are still wilting in the morning. If not, it was just an afternoon slump you witnessed, not a moisture emergency. If the leaves are still wilting in the morning, that's a sign of distress that calls for sending some water through the soaker hose.

Water from a sprinkler will promote leaf diseases on tomatoes. So, use a soaker hose—above the soil, but beneath the mulch—to provide moisture to roots. Soaker hoses are riddled with pores that allow water to seep out along their entire length.

Oxygen

Oxygen in the open spaces between soil particles is critical for root growth. Oxygen also plays a role in sustaining plants at night, until the sugar production of photosynthesis can start up again at daybreak. What's the best way to boost oxygen in the soil? Adding organic matter in the form of manure and/or compost provides food for the earthworms and microbes that change the soil's structure, leaving more room for oxygen. Get a smaller boost for these soil creatures from a layer of organic mulch and application of organic fertilizers.

Sunshine

I saved the best for last.

Picture a whisky barrel being used as a rain barrel. It has an open top and a solid bottom. The sides are composed of seven vertical narrow wooden slats. The slats are held together by several horizontal metal straps.

Your barrel can be full of water *if* all the vertical wooden slats are intact from top to bottom. What happens if one slat is broken off halfway up? Your rain barrel will only ever be half full—even if all the other slats are full length.

Now, imagine that each vertical slat represents something your plant needs: sunlight, nitrogen, phosphorus, potassium, other minor nutrients, oxygen in the soil, and water. If the slat that represents sunlight is broken off at the halfway mark—because the plant is growing in a shady spot, perhaps—then, your tomato plant is only getting half as much sunlight as it needs. The fact that it has access to 100 percent of the other elements it needs—nitrogen, phosphorus, water, etc.—doesn't change that. This is what is known in horticulture as a "limiting factor."

To keep too little sunshine from becoming a limiting factor that negatively affects the growing potential of your tomato plants, always plant them in maximum sunlight. A minimum of eight hours of full sun per day means that plank in the barrel extends all the way to the top.

LET'S *do a* DEEPER DIVE

If you want to know even more about fertility and soils, you couldn't ask for a better guide than *Soil Science for Gardeners: Working with Nature to Build Soil Health* by Robert Pavlis. Pavlis does his homework and knows his stuff.

CHAPTER 4

Tomato Transplants or Seeds—You Decide

ARE YOU GOING TO START YOUR TOMATO PLANTS FROM TRANSPLANTS OR FROM SEEDS? THIS IS REALLY A QUESTION ABOUT TEMPERAMENT, AND YOU WON'T KNOW WHICH ONE YOU FAVOR UNTIL YOU TRY THEM BOTH.

My colleagues and I started a few thousand seedlings when we were farming organic tomatoes in the 1980s. It was satisfying work, especially when the seedlings were ready to graduate from trays and it came time to put those little baby plants in the ground. But by the end of my farming sojourn, my personal feeling was that I had met my lifetime quota of seed starting. Since then, I've bought transplants to start my tomato gardens.

I will direct-seed a few things into the garden, such as cover crops like crimson clover, winter rye, winter peas, and buckwheat. My wife—who also worked on an organic tomato farm in Vermont in her adventurous youth—will direct-seed some resilient vegetables into the garden, such as snap peas, carrots, and greens.

Tomatoes could also be hand-seeded into the garden, but their distaste for cold weather—and our impatience to have that homegrown fruit as soon as possible—means their seedling stage is best spent inside the toasty house.

Fortunately, my tomato teammate Zoe does have the right temperament and excitement about growing tomatoes from seed. The three things you will need to be successful at this are:

1. A selection of varieties that suit your tastes and space
2. A schedule that has your seedlings ready to transplant outdoors soon after the average last frost date in your region
3. A "nursery" that suits the seedlings, your space, and your idiosyncrasies

In this chapter, we'll look at all three of these requirements in great depth.

There are any number of ways to pot up seedlings. Either way, they benefit from a heat mat under them and a grow light above them.

Before we get to the specifics on how and when to plant tomatoes from seeds or transplants, let's look at the many varieties of tomatoes you can plant and discuss the factors that may make a particular type of tomato best for you and your space.

Sizing Up Tomato Plants. Which One Goes Where? And Why?

There are a lot of exciting varieties of tomatoes out there: 'Cherokee Purple', 'Gold Medal', 'Mortgage Lifter', and what seems like a million more. It can be a lot to keep track of and to choose from. But, choosing which tomatoes to grow isn't a test to pass or fail. All the wonderful flavors you read about are as subjective as any person's taste buds. As long as you can plant more than one variety, you'll get to experience a variety of flavors that you—and your fellow gardeners—can grow. Plus, the flavor of any one variety will vary a bit with growing conditions. There is no one right answer for which tomato varieties to grow. You merely have to choose. Yeah. That's all you have to do.

There is one thing I can do, however, to make your choices a little simpler. Let's look at the four forms tomato *plants* can take, and then we'll look at the three forms *the fruits* can take. All the options described should be noted on the plant's tag—or in the catalog write-up—to help you better choose which amazing tomato varieties you'll grow.

TOMATO PLANT FORMS

First, let's think about how big a tomato plant grows and how long a growing season you and your garden can handle.

Plum-shaped paste tomatoes are determinate and can be quickly supported by a cat's cradle trellis (see p. 88).

Do you want a tomato plant that's a nice size for a pot? Or something that can spread out in a big garden space? Then, consider how long or short you want the harvesting season to be. Do you want the fruits ripening all at once for canning, or would you prefer a steady production until frost?

Determinate Tomato Considerations

There's something I love about the definition of the word "determinate":

adjective

1. Having defined limits
2. Definitely settled
3. Conclusively determined

(Thank you, *Merriam-Webster Dictionary*.)

As "determinate" tomatoes, these varieties have a "defined limit" of about 4 to 5 feet (1.21 to 1.52 m) high, making them good companions with the cat's cradle trellising method detailed on page 88, or a 4-foot (1.21 m)-tall, square tomato cage, as shown on page 91. They also pack a long season's worth of ripe fruits into a "conclusively determined" period of about eight weeks.

Many commercial tomato growers have started naming their determinate tomatoes with the more user-friendly term "bush." As in 'Early Girl Bush'. Often, the terms "bush" and "determinate" can be used interchangeably.

Determinate paste tomatoes are great for homemade pizza sauce.

Who Would Choose *Determinate* Tomatoes?

DETERMINATES ARE YOUR FRIEND IF YOU:

- Want to harvest and/or preserve a lot of tomatoes in a short period
- Have a short growing season
- Have a long growing season, but want to pull out your tomatoes in late summer to make room for a fall crop of greens, root crops, or crucifers
- Want to use only the easier-to-store cat's cradle trellis system (see p. 88) as compared to bulkier tomato cages

OTHER CONSIDERATIONS FOR CHOOSING DETERMINATE TOMATOES:

- As seedlings, they look the same as indeterminate tomato plants so manage your labels well.
- A flowering cluster at the top of the plant signifies the final height and last fruits of that season. Lopping off the top will reduce the number of flowers and fruits with no other benefit.
- The flower clusters develop at the end of flowering branches so any pruning dials down the number of fruits produced.
- That also goes for pruning suckers—doing so will reduce the number of fruits, rather than boost them.
- Because a finite amount of leaves have to photosynthesize and produce sugars for a more concentrated weight of fruits, determinates' flavors—with a few exceptions—aren't always as robust as the taller, longer-season indeterminate tomatoes.

Indeterminate Tomato Considerations

Now that we know the definition of determinate, we can more easily understand the meaning of "indeterminate." According to *Merriam-Wester*: "Characterized by growth in which the main stem continues to elongate indefinitely without being limited by a terminal inflorescence."

Growing indeterminate tomatoes helps make clear that tomatoes of any kind are not annuals. They are tender perennial plants that will keep growing leaves, flowers, and fruit until the first hard, killing frost of autumn. In climates as warm and winter-free as the Peruvian and Chilean coasts, where tomatoes were designed by evolution, they will continuously grow and produce fruit for several years before aging out.

As tomato plants "elongating indefinitely," indeterminates will get as tall as you let them until the first hard frost kills them. That makes them good candidates for tomato cages. Indeterminate tomatoes can grow taller than the 5-foot (1.52 m) height of these cages, but I merely let them grow out and over the top and then drape down the sides unassisted. The important thing is that they are not on the ground where vermin and diseases can more easily reach them. Unless you're in a winter-free climate, frost will kill the plants before they reach the ground anyway.

A mature beefsteak tomato inside a tomato tower. Lower leaves have been removed to keep soil diseases like early blight from spreading. Even partially ripe tomatoes—like this one—can be harvested before a big rain, so they don't split. They can be ripened indoors without loss of flavor.

Who Would Choose *Indeterminate* Tomato Plants?

INDETERMINATES ARE YOUR FRIEND IF YOU:

- Want to have slicers, cherry, and grape tomatoes until frost
- Crave a particular heirloom variety flavor
- Can manage plants that could get taller than you

OTHER CONSIDERATIONS FOR CHOOSING INDETERMINATE TOMATOES:

- As seedlings they look the same as determinate plants so manage your labels well.
- Flowers will not appear at the top of the plant, but only from the side branches so chopping off the top of the plant won't keep new flowers and fruit from appearing.
- With a larger proportion of leaves to fruits, these plants can photosynthesize more flavor than many of the determinate varieties.
- Removing suckers—of which there will be many—doesn't improve the numbers of fruits or the flavors. If you're in a humid climate, removing suckers *may* reduce the amount of leaf-borne diseases. But, hopefully, you're addressing that with mulch and removing lower leaves already.

Dwarf Tomato Considerations

A dwarf tomato has a thick stalk and short enough stature—about 2 to 3 feet (60.96 to 91.44 cm)—to sometimes grow without a trellis. They're shorter than determinates, but the higher ratio of leaves to fruit means they can produce tastier fruit. The flavor of many dwarfs is comparable to some heirlooms, but the fruit size is often no more than a ½ pound (225 g), while indeterminate fruits can often be a full 1 pound (454 g).

Plant breeders had created dwarf varieties in the 1890s. This was even before breeders developed the better known determinate varieties, like 'Roma', in the 1920s. But, until the folks with the Dwarf Tomato Breeding Project in 2006 began developing new varieties, there were few dwarf varieties available, and they were hard to find. By 2019, the Dwarf Tomato Project offered more than one hundred new dwarf varieties in many colors, sizes, and shapes, many of which are now available in garden catalogues.

Dwarfs grow well in 5-gallon (18.92 L)—or larger—containers on a sunny deck or patio as well as in straw bales. And because these plants are shorter and their stems are much stockier than indeterminates, they may do well without a trellis, even in straw bales.

Who Would Choose *Dwarf* Varieties?

DWARF VARIETIES ARE YOUR FRIEND IF YOU:

- Have enough space for determinates but want the flavor of heirloom indeterminates
- Have enough space for determinates, but want to forgo trellises
- Grow in straw bales and pots

OTHER CONSIDERATIONS FOR CHOOSING DWARF VARIETIES:

- The seedlings are recognizable from their very thick, stocky, and shorter stems that provide more support than other types.
- The leaves are also noticeable for being crinkly and dark green to bluish green.

Micro Dwarf Considerations

These gems will grow in a small 1-gallon (3.78 L) container. That's the same size that many perennial flowers and even houseplants grow in at the garden center. They get no more than 1 foot (30.48 cm) or so tall. The flowers, leaves, and fruit are packed together densely. With supplemental light and a little help with pollination, they can even grow and produce fruit indoors.

Who Would Choose *Micro Dwarf* Varieties?

MICRO DWARFS ARE YOUR FRIEND IF YOU:

- Don't have space outdoors for a tomato garden
- Enjoy tiny fruits
- Don't want to fuss with a trellis system

OTHER CONSIDERATIONS FOR CHOOSING MICRO DWARF VARIETIES:

- These varieties are also great for hanging baskets, living walls, deck planters, and other unusual growing sites.
- The tiny fruits can be quite sweet but some varieties have thicker skins.

A micro-dwarf variety grows in a cowgirl boot repurposed as a plant container at the entrance to the straw bale patio snack shack. Sweet potato leaves promise a late season crop too.

Zoe grows 'Rosella Purple' dwarf slicing tomatoes.

TOMATO FRUIT FORMS

Tomato fruits come in three basic forms: slicers, paste, and cherry and grape.

Slicers

A grilled cheese and tomato sandwich topped with edible portulaca greens from the garden's living mulch plants.

A "slicer" is a tomato that produces fruit in the ½- to 1-pound (225 to 454 g) or better range. Slicers make a great sandwich addition and can also be chopped for salads and salsas. Many popular indeterminate heirlooms, like 'Cherokee Purple' and 'German Johnson', produce massive 1-pound (454 g) tomatoes. Also, there are commercial hybrid slicers, like 'Better Boy', which produce in the ½ pound (225 g) or better range.

There are determinate plants that grow great, round, juicy slicing tomatoes too. These are commercial hybrids like 'Celebrity', 'Mountain Pride', and others that are favored by farmers because they produce over a short season of about two months.

Popular dwarf slicer varieties include 'Rosella Purple', 'Summertime Green', and 'Dwarf Mr. Snow'. These fruits have the color, taste, and size equivalence to ½-pound (225 g) heirloom tomatoes, but the plants grow to no more than half the size.

Slicers come in a variety of vibrant colors that can ramp up the look of any tomato dish.

Just *How Green* Are Your Fried Green Tomatoes?

When Mrs. Calhoun, my white-haired neighbor across the street, saw my tomato plants covered with immature fruit, she flagged me down and said: "If I can have some of those, I'll fix you up some fried green tomatoes." I was 22 years old, living in my hometown of Beaufort, SC. My Wisconsin-born mother had mastered southern fried chicken to please my father, but she had never prepared fried green tomatoes in our culturally mixed house. But I was a growing boy, living on my own and someone was offering to cook me some food. What could I say? "Sure, lets pick 'em right now."

Mrs. Calhoun chose four of the nicest green tomatoes and one that was smaller, handed them to me to carry and said, "Come on in the house." In her kitchen she set up one shallow bowl of whole milk and another similar bowl of flour with salt and pepper.

She sliced a couple of tomatoes, saying, "We can eat the rest tomorrow." She then showed me the difference between what she called an "unripe tomato" and a "green-ripe tomato." When she sliced the smaller one that she called unripe, she showed me that many of the seeds were soft enough that they had been sliced in half by her knife. But the green-ripe tomatoes had seeds that were mature enough and solid enough that they slid around the edge of the knife and remained intact. "If the seeds are whole," she said, "then the tomato is green-ripe and will taste delicious after we fry 'em up. But if you use the unripe tomatoes, the flavor will be disappointing."

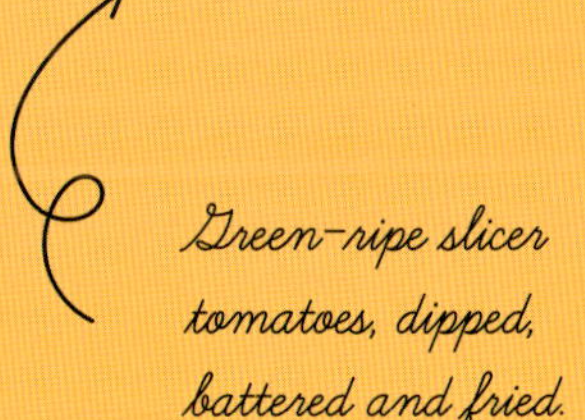

Green-ripe slicer tomatoes, dipped, battered and fried.

Mrs. Calhoun put each slice in the milk just long enough to wet it good on both sides. She then laid each dripping piece in the flour mixture on one side then the other to coat them all over. As each slice was ready, she gently positioned it in a wide, cast iron pan holding a thin, glistening, bubbling layer of melted butter. When both sides were browned, she set each slice on a paper towel covering a plate.

With all the slices done, Mrs. Calhoun sat us down at her kitchen table, each with a plate of her fried green tomatoes, so we could enjoy our afternoon snack. I followed her lead, cutting the tomato slices with a knife and forking pieces into my mouth. They had a crispy coating on the outside, juicy flesh on the inside and a flavor that I have only ever been able to describe as "delicious, but not like a tomato at all."—I had found a new favorite food.

Paste

Some popular determinate varieties, such as 'San Marzano', 'Amish Paste', 'Polish Paste' and 'Roma', are called "paste" tomatoes. All paste varieties produce a pear-ish-shaped fruit that's not as juicy as a big slicing tomato. Bred to hold less water so they cook down more easily, paste tomatoes make a great sauce. And that sauce is often canned to extend the tomato-enjoying season through winter.

For all of us who like to color outside the lines, paste tomatoes also are fine sliced on a sandwich—if you want less tomato juice running down your chin. And there's no law against using them in a salad or salsa either.

Cherries and Grapes

The difference between cherry and grape tomatoes? Cherries are larger and rounder, like a cherry. Grapes are smaller and more oval, like a grape. Got it? Depending on the variety, the sweetness and flavor can vary. Grape tomato skins are thicker, with less flesh, so they don't squirt as much as a cherry tomato when you bite into it. The thicker skin also means they last longer on the counter than cherry tomatoes. Both are suited to a bit of grilling to enhance their flavor.

A selection of fruits from indeterminate heirlooms, dwarfs, and cherry tomatoes.

Planting Schedule

Regardless of which types of tomatoes and fruit forms you choose to grow, the next step is to get the timing and planting methods right. Remember, tomato plants are short-lived perennial sprawling vines—as opposed to climbing vines—that evolved on the coasts of Peru and Chile without being exposed to freezing temperatures. As a result, they have no resistance to freezing spring weather or freezing fall weather and can be killed when it's that cold.

Also, their cells can be injured and/or their growth slowed when the plants are exposed to temperatures just above freezing. In other words, they grow, flower, and fruit best when the temps are around the same temperatures we humans like, say from 60°F to 85°F (15.5°C to 26.6°C).

In most regions, your local garden center or agricultural extension office will be able to tell you the last average frost date. You can also conduct an internet search with your zip code and "average last frost date" to find this information. Understand this isn't a guarantee of the last day of frost, but the *average day* of the last frost where you live, meaning your region may experience freezing tomato-killing temperatures up to two weeks after that. Do not plant tomatoes outdoors until after this date has passed and your long-term forecast shows no threat of nighttime frost.

Transplants or Seeds?

I still remember one of my tomato farm colleagues kicking a watering can across his backyard. He'd come out to water our thousands of seedlings and discovered he'd waited too long. About a hundred seedlings, in a couple of flats, were dried up and dead. Fortunately, we had started more than we needed so all was not lost. But the lesson to all of us was clear. Seedlings are a bit like human toddlers. They are totally charming, but turn your back too long and they will have shoved a bean up their nose by the time you get back to them.

Let me cut to the chase on this subject. I enjoyed starting and nurturing thousands of tomato seedlings years ago for the co-op's organic tomato farm. But sometimes, something that's new and exciting can quickly lose its luster. Now, I buy transplants. At that stage, they are less like toddlers and more like tweens. You can turn your back on them for a few days and there's no bean up their nose when you come back. They don't need a babysitter all the time, and they seem to enjoy your company.

Are you trying to decide whether to start with transplants or seeds? At the start of this chapter, I declared it a question of temperament, a "know thyself" kind of thing. What do I mean by that?

Well, being a seed starter is a calling. There are gardeners who love, love, love to get seed catalogues, make lists, order supplies, handle damp potting mixes, compare and contrast potting mixes, chat online or IRL with fellow seed starters, monitor conditions, agonize about conditions, pot up and pot up again, take seedlings outside when it's warm, bring seedlings inside when it's cold, take them back outside again, mourn the occasional losses, share the extras with friends, and maybe cover some of their costs by selling much-sought-after seedlings.

How do you know if you're called to be a seed-starting-type of gardener? You can really only know by trying. It's much like finding one's vocational calling. It's not an intellectual choice. It's an emotional one. You won't find your vocational calling by looking in the college course catalog. Finding one's vocational calling and determining whether or not seed starting is your thing is like going to a party and meeting someone on whom you quickly develop a crush. It's that *sense* of a crush (on that person—or your seedlings) that tells you you've found your calling. Once you set up your seed-starting station, you will realize whether you're called to nurture seeds and seedlings. Or not. Again, it's not an intellectual decision. It's an emotional one. Trust your feelings. So, let's start by taking a look at both starting from transplants and starting from seeds to help you decide which is best for you.

TRANSPLANTS

Many garden centers (especially big-box retailers) will start carrying tomato transplants well before it's time to put them in the ground. Before nightfall, they will pull their many flats of baby tomatoes back into their building and then move them back out again in the morning to tempt you. Independent garden centers and university agricultural extension offices—and experienced gardeners—all know that the conventional time to put transplants in the ground is *two weeks after* the average last frost date of spring. Why, then? Tomatoes evolved in a frost-free region so they don't have resistance to conditions much below freezing. Even air temps in the 40s and 50s Fahrenheit (4 to 10 degrees Celsius) will take the wind out of their sails, even if they don't kill them. Plus, tomatoes expect the soil to be warm enough to keep the roots growing strong. Cool soil puts the brakes on the speed of healthy growth even worse than cool air does. Tomatoes—like all plants—are similar to cold-blooded animals like turtles and frogs. The speed of their metabolism and growth is determined by the temperature of the air and soil: warm = fast; cold = slow.

Many gardeners play it safe by not being in a rush to plant tomatoes outdoors. Setting them in the ground as late as three or four weeks *after* the average last frost date is no crime. Tomatoes planted in warmer conditions will ripen their first fruit not much later than their early planted cohorts, because they've been able to keep up a steadier pace of faster growth in warmer soil.

You can hedge your bets on the temperature in one of two ways: use black plastic mulch (see p. 155), which transfers the sun's warmth to the soil, and/or use baby greenhouses (see p. 157) while the transplants are short to retain the soil's warmth overnight.

My local seed-starting guru, Zoe, labels containers for her baby tomatoes.

How to Plant a Transplant

Once you've chosen the best time to plant and your beds are prepped, use a trowel to dig a hole for your transplants. Then, toss two handfuls of organic fertilizer into the hole. Set the roots deep in the soil, so some of the stem is also underground. Because tomatoes are much like a vine, they will form roots off the stem. It doesn't matter if the organic fertilizer touches the stem: It's just leftover plant parts (like cottonseed meal), animal parts (like bonemeal and bloodmeal) and probably some soil as a filler. As you know now, conventional synthetic chemical fertilizer, however, is functionally a salt so don't let it come in contact with any part of the tomato plant. The saltiness of a synthetic fertilizer will draw moisture out of the plant and burn it. Possibly killing it.

Use the trowel or your hands to replace the soil around the stem and lightly push it down. A slosh of water on the soil, over the roots, from a watering can is a good finishing touch for the planting. Surround the base of the transplant with one-half of a toilet paper roll if cutworms are a problem in your area.

When all transplants are in, place a soaker hose alongside each plant. Don't let the soaker hose touch the stems or it will cause them to rot. Then, cover the soaker hose and soil with a protective layer of 1 to 2 inches (2.54 to 5 cm) of organic mulch to slow the evaporation of soil moisture and to keep rain from bouncing soil fungus onto the lower leaves.

Starting lots of different vegetables from seed, including tomatoes, saves money and gives you more choices.

STARTING FROM SEED

By starting from seed, you'll have more choices on which tomatoes to plant. Many more choices. Hundreds of choices. Don't let yourself get overwhelmed. Look at the beds you have in mind for planting tomatoes and make a list:

How many paste tomatoes for sauce?

How many cherry tomatoes for snacking?

How many micro dwarfs for growing in containers?

How many slicers?

Will the slicers be indeterminate, determinate, or dwarfs?

Pen and paper with columns will help you stay on top of your choices. And when the season is over, and you wish you'd grown something different, find relief in planning for next year. On paper, at least.

Your Seed-Starting Setup

Before the seeds arrive, you'll want your seed-starting station to be ready.

Timing

Plan to put seeds in their pots no sooner than six weeks before the last average frost date. That means at eight weeks, it will be two weeks past the last average frost date, a traditionally safe time to put tomatoes in the ground. Starting seeds a bit later is fine. Starting seeds earlier means your transplants may be rootbound and leggy by the time conditions outside are warm enough.

Heat Mat

These waterproof mats, with an electric cord, lie under your flats to keep seeds, soil, and plants exposed to an even, warm temperature. A heat mat will get your seeds germinating faster and your plants growing more steadily at a temperature that's much warmer than you would want to keep your house thermostat. Since warm air rises, you're not only warming the soil and roots, but you're also creating a toasty bubble of warm air for the leaves and stems to bask in. Like swaddling a baby.

Shop Lights

You *can* start seedlings with just a sunny window, but they will be soooo hungry for more light that they will stretch toward that window. To keep them from falling over, you would need to rotate the flats once a day, and they will still get much thinner and taller than is ideal. It's better to invest in a shop light and mount it just above the flats. You don't need expensive horticultural fluorescent tubes, although the adjustable stands they come with are helpful. Common shop lights from a hardware store hold two tubes, so install one tube that's labeled "warm white" and one tube that's labeled "cool white" to get the broad spectrum of light that plants want. Turn them off at night, because, like us, tomato seedlings need the downtime of darkness to grow properly. Adding a timer is an unnecessary expense, unless you're not available to turn on the light in the morning and off in the evening for some reason. This is also a good schedule for checking on their ever-changing conditions. Here, again, I find it helps to think of seedlings as being like human toddlers—things happen quickly in their young lives.

You can put more than one seed in each container to optimize space.

A flat of tomato starts that have all germinated.

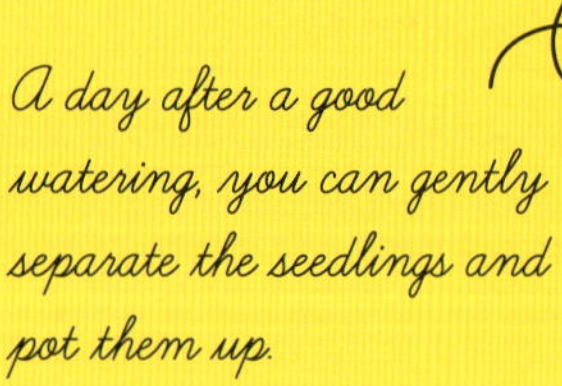

A day after a good watering, you can gently separate the seedlings and pot them up.

Containers

Your best bet is to use a two-part system of sturdy, reusable trays. The first part is a *cell tray* that holds potting soil and can hold as many as fifty seedlings in fifty individual cells that allow excess water to drain out. That cell tray then rests inside a *seed tray* which has no drain holes and, therefore, holds water: a bit like a saucer beneath a houseplant's pot. This kind of setup allows for bottom watering, meaning you don't risk wetting the leaves or waterlogging the soil by watering from the top. By pouring water into the lower seed tray, the bottoms of all the cells are exposed to water and can absorb the moisture into the potting soil to keep the tomato seedlings from drying out.

Alternative: No Containers

The option we went with on the farm was to use the solid seed trays lined with a layer of cloth, like an old hand towel, to absorb and distribute excess water. On top of that cloth, we placed what are called "soil blocks" made by using a device called a soil blocker. Soil blockers are rust-proof metal contraptions that compress potting soil into blocks, allowing you to grow each seedling in it's own compartment without needing any pots. We put the potting soil in a wheelbarrow, wet it down until it was soggy, then pressed the soil blocker into the soggy potting soil until the blocks were full. We then set the loaded soil blocker into the tray and squeezed the handle to release the blocks. Some people start their tomato seeds in tiny blocks—or cells—and then, after a month, move them to a bigger container or a larger block. We used the 2-inch (5 cm) blocks, started the seeds in them, and they did fine. I think with tomatoes, it's worthwhile to skip the up-potting step. Everyone has too much to do already.

Alternate containers: If you want to avoid plastic containers or if some of your seedlings will be gifts, you can make charming pots of bamboo, burlap, and string.

Moisture

Even with the right cell and seed trays as described previously, you still need to be a good moisture manager: not too much and not too little. The key to managing moisture, beyond bottom watering, is to be checking your plants once or twice a day to see how they're faring. Given that humans are not perfect, I think it's important to decide which side to err on. With plants like tomatoes, they're better off a little too dry rather than a little too wet. When they're too wet, you risk what's called "damping off," which is a fungal disease that attacks the roots of seedlings. As we've discussed in regards to the origins of tomato plants, they evolved in a relatively dry climate. That means they didn't evolve the kind of immune system needed to fight off fungi that thrive in damp environments. To grow, they need a steady amount of moisture, but too much—often indicated by a constantly damp potting soil surface—invites disaster. If the soil gets dry enough for the leaves to wilt just a bit, that is better than being too wet and having the whole plant die when its root system fails. With tomatoes and other plants, I've noticed a stage that comes before wilting. Plants go from looking perky to looking less perky. You'll notice this if you're keeping a regular eye on conditions. If the leaves are less perky and the potting soil surface looks dry, then give them a dose of water before the day is out. It's not an immediate crisis. All live things do better if they are exposed to some stress. It makes them stronger.

Potting Soil

I believe we live in a "you get what you pay for" world. So, don't skimp on your potting soil. Okay . . . it's not really soil. The more accurate term is "potting media," or "potting mix," but I've never heard an actual gardener use those terms. So, since most bags of potting soil don't show a detailed ingredients list, buy the most expensive one or at least chase down a bag of professional-grade potting soil. Why? It's really about the oxygen. Too little attention is paid to oxygen as a plant nutrient. Yes, the leaves do absorb carbon dioxide, but the roots are hungry—thirsty?—for oxygen. A good quality, professional potting soil is created so it holds lots of water and oxygen in its pores.

How to Plant Tomato Seeds

Tomato seeds are tiny. What to do if you have big fingers? Get out the chopsticks. Or at least one chopstick. Once your cells or soil blocks are ready, dump your seeds on a wide plate, put some water in a small bowl, then dip the end of your chopstick in the water. Use the damp end of the chopstick to touch a single seed and watch water adhesion hold the seed until you can carefully place it in the center of the potting soil in each cell or block. Wipe the seed against the potting soil to dislodge it from the chopstick, then lightly mash it into the soil with the tip of a finger. Repeat.

Tending Your Seedlings

Care for your seedlings properly and they'll be ready to transplant into the garden when the time is right.

Air Movement

When seedlings move in the breeze, it stimulates them to grow shorter, sturdier stems. You can recreate this benefit for seedlings growing indoors by leaving the ceiling fan or a box fan on the "low" setting for a few hours each day. No need to aim the box fan right at them. Aim it at any other part of the room. You just need enough airflow to see your seedlings waving at you when you enter the room.

Hardening Off

When the weather outside has warmed to between 60°F and 70°F (15.5°C and 21.1°C)—but before the average last frost date—bring the seedlings outdoors to a shady spot to get them exposed to the actual conditions they'll face once you plant them. This is called "hardening off." If night temps are below 60°F (15.5°C), bring them back indoors. If night temps are above 60°F (15.5°C), you can leave them out and move them to a slightly sunnier spot each day or so to get them used to their new growing conditions.

Planting Day

If you're two weeks or more past the average last frost day, you can gently remove the root ball from the trays, or separate each individual soil block if you opted for soil blocking, and plant your seedlings out in the garden. Revisit Option One (see p. 76) for details on this step.

Next, it's time to dig into the subject of supporting your growing tomato plants with various trellising, staking, and caging systems.

LET'S *do a* DEEPER DIVE

The New Seed-Starters Handbook by Nancy Bubel covers seed starting for tomatoes, other vegetables, fruits, flowers, trees, and herbs.

Epic Tomatoes by my colleague Craig LeHoullier goes into great detail on enough varieties to fill a book.

www.theseasonalhomestead.com has the best coverage on starting seeds in soil blocks, even sharing that they've been used for 2,000 years.

CHAPTER 5

Tomato Structures: DIY Tomato Towers, Cat's Cradles, and Tomato Houses

THE TOMATO PLANT IS THE QUEEN OF THE GARDEN, AND SHE DESERVES A PROPER CASTLE TO HELP HER RISE ABOVE HER ENEMIES: DAMP SOIL, FUNGI, AND GROUND-DWELLING VERMIN. TOMATO PLANTS NEED A PROPER TRELLIS TO HOLD THEIR LEAVES AND FRUIT ABOVE THE GROUND. IN THIS CHAPTER, WE EXPLORE VARIOUS METHODS FOR SUPPORTING YOUR GROWING PLANTS BASED ON THE TYPE OF TOMATO YOU ARE GROWING.

As a quick recap, there are two kinds of tomato plants—indeterminate and determinate—and they each need their own kind of trellis.

First, *indeterminate* plants continue to grow taller until killed by frost, meaning their height is undetermined. So they benefit from a tall tomato tower.

Second, what's commonly called a Florida weave (see p. 86), works best with *determinate* tomatoes. These are tomato varieties that stop growing at 4 or 5 feet (1.21 or 1.52 m) high by midsummer.

Common—and Often Inefficient—Trellising Techniques

Store-bought "tomato cages" are no more than 3 feet (91.44 cm) high once their wire legs are pushed into the ground. They are not nearly tall enough or wide enough for most tomato plants. They can fall to the ground easily under the weight of even a below-average tomato plant. Most beginner tomato gardeners only use these once.

Fencing wire can be fashioned by the home gardener into a better cage than the store-bought ones, but the openings are too small for harvesting any decent-size fruit. They're also expensive and covered in zinc, and so have a high carbon footprint.

Perhaps the oldest technique—tying each branch to a wooden stake, like so many floppy hostages—takes a great deal of time. Plus, the wooden stakes eventually rot away and must be replaced.

What's a *twenty-first-century* tomato gardener to do? *Well,* let me tell you.

LET'S *do a* PROJECT

A Cat's Cradle to Support Shorter, Determinate Queens of the Garden

I WORKED AS AN INTEGRATED PEST MANAGEMENT SCOUT ON TOMATO FARMS IN SOUTH CAROLINA IN THE 80s, HELPING REDUCE THE AMOUNT OF CHEMICALS SPRAYED ON FIELDS.

I still remember watching the arm of a farm worker go up and down like a drum major's wand. He was walking down a row of tomato plants and looping twine over each stake to create what's called a "Florida weave." This simple technique keeps the tomato plants from lolling on the ground where their fruit could rot. Also, by raising the tomato plants higher, their fruits are easier to pick.

This technique is rarely seen in backyard gardens. But, because it's such a time saver, it deserves to be. In our garden, we can tie up fifteen 'San Marzano' paste tomato plants once a week in less than three minutes using this method. How long would it take you to tie up fifteen plants and their branches to single stakes? A lot longer than that, I can assure you.

Since most tomato gardeners are not in Florida, I propose we give this technique a catchier name. I coined one for an article in *Organic Gardening* magazine: "cat's cradle."

Why?

Because many of us grew up knowing how to play the string game of the same name that's as simple as this trellising method. And because once the trellis is complete, it somewhat resembles the beauty of a cat's cradle string structure. And lastly, this trellis does in fact "cradle" the tomato plants.

SUPPLIES NEEDED

- One more stick of rebar than you have plants to trellis: so, six rebars for five plants; each stick should be 5 to 7 feet (1.52 to 2.13 m) long
- A roll of jute twine
- A hammer for pounding the rebar into the ground
- A level if you're not confident about setting the rebar stakes perpendicular-ish by hand
- A metal cutting blade and saw if you're cutting longer rebar to length

THE METHOD

Here's how to make your own cat's cradle:

1. After planting a row—of any length—of tomato transplants, drive wooden stakes—or, my preferred material, indestructible metal rebar—8 to 12 inches (20.32 to 30.48 cm) into the ground between each plant and at the end of each row. If set deep enough, some gardeners may be able to get away with setting one stake after each *pair* of tomato plants for a big savings in time and money.
2. Tie your natural and compostable jute twine to the *first* stake, setting it a little lower than the highest leaves of your tomato transplants.
3. Loop the twine around each *intermediate* stake, at a height that cradles each plant, until you get to the end.
4. Loop the twine two times around the *end* stake for stability.
5. Work your way back up the row, adjusting the height of the twine so it continues to cradle the tomato plants to prevent them from flopping over.
6. Every five to seven days, tie and loop another, higher, length of twine around each stake to cradle the new growth on each plant.
7. Tuck wayward branches into the cradle as you go.
8. At the end of the tomato harvest season, snip each strand of jute twine with a knife or scissors and let it fall to the ground to decompose. Yank out the plants for composting and remove the rebar stakes for off-season storage. Prepare the ground for next season's planting.

The advantages to the home gardener are that the cat's cradle takes *much* less time to implement than tying each plant and its flailing branches to an individual stake. A cat's cradle also better contains all the loose side branches of the tomato plant in its woven web. And removal at the end of the season goes more quickly.

Vertical rebar and horizontal twine are the fastest way to trellis determinate tomatoes, including plum and dwarf tomatoes.

LET'S
do a
PROJECT

A Cat's Cradle *Alternative* for Determinate Tomato Plants

MAYBE PLANTING YOUR DETERMINATE TOMATO PLANTS IN A STRAIGHT ROW FOR A CAT'S CRADLE TRELLIS ISN'T AN OPTION FOR THE SHAPE AND SIZE OF YOUR BEDS. IN THAT CASE, HEAD FOR THE HARDWARE STORE AND BUY SINGLE PANELS OF REMESH, ONE FOR EACH PLANT.

They have the same rusty patina and 6 × 6-inch (15.24 × 15.24 cm) openings as the big rolls of remesh, but the sheets are only 5 feet (1.52 m) tall and 7 feet (2.13 m) long. Because they are manufactured as flat panels rather than produced as a roll, they are VERY HARD to curve into a tube, as we've done easily with rolls of remesh. But the sheets can be pretty easily shaped into a square footprint that can slip over your transplants and be stabilized just like a round tomato tower with a rebar stake.

THE GENERAL IDEA

These panels let you make a 4-foot (1.21 m)-tall tomato tower that will last forever and can be stored outdoors. The openings are wide enough that you can slip a 1-pound (454 g) ripe tomato through them. The footprint is square instead of round. Each side is 21 inches (53.34 cm) wide. Times four sides, adds up to the 84 inches or 7 feet (2.13 m) of the long side of these panels. The ends are held together by what are called "hog rings," tie wire, zip ties, or twine.

SUPPLIES NEEDED

- One 3½ × 7-foot (1.21 × 2.13 m) panel of remesh per determinate tomato plant
- Chalk for marking
- One 2 × 4 (38 × 90 mm) board, at least 4 feet (1.21 m) long
- 9 hog rings (or tie wire, zip ties, or twine) per remesh panel
- Hog ring pliers
- 4-foot (1.21 m) lengths of rebar to anchor the finished towers
- Tie wire, zip ties, or twine to secure the anchor rebar to the towers

THE METHOD

Start with One Bend

Lay your 3½ × 7-foot (1.21 × 2.13 m) panel of remesh on the ground. Starting with a 3½-foot (1.21 m) end, use chalk to mark out 21-inch (53.34 cm) increments at the top and bottom of the 7-foot (2.13 m) sides. You will have three marks on each side. You should be able to see how your marks have defined the four sides of your square tower.

Lay a 2 × 4 (38 × 90 mm) that's at least 4 feet (1.21 m) long on the panel, aligning its edge with the first chalk marks on the two long edges. While standing on the 2 × 4 (38 × 90 mm), reach over and grab the 4-foot (1.21 m) end and lift it up until it's as close to vertical as you can get it. The important thing is starting a bend. Once each 6 × 6-inch (15.24 × 15.24 cm) square that's bisected by your 2 × 4 (38 × 90 mm) has that crease, you will be able to shape the tower the way you want it.

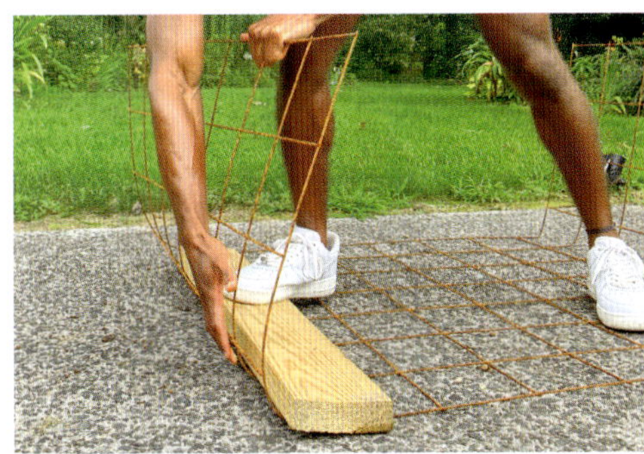

Dyson uses a length of wood to bend a single sheet of 42" × 84" (1 m × 2 m) remesh into a tomato tower for a dwarf or plum tomato.

Make the Second and Third Bend to Create Four Sides

Move the edge of the 2 × 4 (38 × 90 mm) to the chalk marks on the opposite end. Repeat the steps just listed. At this point you'll have the 42-inch (106.68 cm) middle of the panel lying flat on the ground, with chalk marks at its halfway point. At each end, you'll have a 21-inch (53.34 cm)-tall section aiming up into the air. The panel will vaguely resemble a wide letter *U*. Place the edge of the 2 × 4 (38 × 90 mm) along the two center chalk marks. Kneel on the 2 × 4 (38 × 90 mm) and bend the panel against the board as far as you comfortably can. It doesn't matter if you're able to bring the two 3½-foot (1.21 m) edges of the panel together at this point.

A finished, square tomato tower. Each of the four sides will be composed of three and a half squares of remesh. Where they join can be secured with either twine, tie-wire, zip ties, or something similar.

Make a Square Tube

Remove the 2 × 4 (38 × 90 mm). Stand over the bent panel and push down on each side to get each corner closer to a 90-degree angle. Gentle leaning and some light bouncing are my preferred techniques. Do this for each corner. The angles and the lines don't have to be perfect. The important thing is getting the two 3½-foot (1.21 m) edges to meet.

Rebar's Advantages Over Wooden Stakes

1. The rebar stakes' tops don't splinter from being hammered into the ground.
2. The rebar stakes are narrower so they drive into hard soil more easily.
3. Rebar's dark rusty color is more attractive in the garden than wooden stakes covered with mildew.
4. The rebar stakes can be pulled out of the ground at the end of the season without getting splinters in your hand.
5. The rebar stakes can be stored outside without rotting as the initial layer of rust protects the rest of the steel.
6. Rebar stakes last longer so they save money in the long run.

Rebar at most big-box stores can be too tall or too short for a cat's cradle trellis. If you have a power saw with a metal cutting blade, you can cut the 10-foot (3 m) pieces of rebar down to 6 or 7 feet long (1.82 or 2.13 m). Or if there's a lumberyard nearby, they have this tool for cutting rebar to length. I have them cut a 20-foot (6 m) length into three lengths of 6 feet and 8 inches (203.2 cm).

Get the Right Rebar

1. For determinate tomato plants in a cat's cradle, the stakes need to rise at least 4 feet (1.21 m) above the ground.
2. At big-box stores, the common sizes for rebar are 4-foot (1.21 m) about $5, and 10-foot (3 m) about $8.
3. The rebar's thickness may be ⅜ inch or ½ inch (0.95 cm or 1.25 cm) and both are just as good for our purposes, but the skinnier ones cost less.
4. The 4-foot (1.21 m) rebar will be little more than 3 feet (91.44 cm) high once you've pounded it into the ground and will work well enough as a cat's cradle: You can just let the tallest part of the plants hang down.
5. The 10-foot (3 m) lengths are too long to hammer into the ground, but can be cut in half with a metal cutting blade on a circular saw or grinder. An employee at a big-box store may be able to do this for you for a nominal charge. You'll get two 5-foot (1.52 m)-tall pieces that will be 4 feet (1.21 m) tall once you pound them into the ground. And you'll have saved money over buying the shorter ones. This is the best option for most people.
6. If you're tall, I think the best bet is to find a lumberyard with 20-foot (6 m)-long rebar, about $12, that they will cut into three pieces for a nominal charge. Ask for three lengths of 80 inches (203.2 cm; equivalent to 6 feet 8 inches in length). This is plenty for staking a cat's cradle. And stakes this long will also be useful for other garden, homestead, and farm tasks in the off-season.

LET'S
do a
PROJECT

The World's Best Tomato Towers for Indeterminate Tomatoes

THE PERFECT ROYAL CHAMBER FOR THE QUEEN OF THE GARDEN SHOULD:

1. Be tall enough to keep the royal fruit off the ground
2. Have openings big enough to extract ripe, 1-pound (454 g) 'maters
3. Look good
4. Last indefinitely
5. Survive the elements when stored outdoors
6. Be a set-it-and-forget-it kind of system

For forty years, I've been using long-lasting tomato towers—often also called tomato cages (but it's not like we're containing a wild animal)—made from concrete reinforcing mesh. This is the metal fencelike material laid down by concrete workers before pouring a sidewalk or floor. It's the right size, shape, and material to strengthen concrete. But it's also the best material for making tomato towers for indeterminate tomato plants. Who'da thunk?

This mesh resembles fence wire as it comes in a roll, but it's much stronger and meets all of our royal requirements, including:

1. It makes a tower that's 5 feet (1.52 m) tall (taller plants can hang down the outside of the tower and still grow great fruit).
2. The 6 × 6-inch (15.24 × 15.24 cm) openings make it easy to harvest 1-pound (454 g) tomatoes.
3. The permanent patina of rust it acquires looks earthy.
4. They can be stacked and stored outdoors in the off-season without any deterioration so they needn't take up space in the garage, basement, or barn.
5. My towers are as sound today as they were when I made them decades ago.

You can buy a roll of concrete mesh at most building supply stores. One roll will cost about $180, weigh roughly 100 pounds (45.35 kg), and make at least twenty-five towers. At that rate, they cost $7 apiece—about the same as the flimsy, short, ineffective tomato cages on the market. Since few gardeners need twenty-five towers, building them is a perfect project for:

A roll of 5-foot-tall (1.5 m) remesh is heavy, but it can be easily moved with a hand truck.

1. Gardeners with lots of gardener friends with whom to share them
2. Community garden members
3. A family farmer with a CSA
4. A gardener who—as I do—wants to sell surplus towers to other gardeners as a side gig (For each of the last five springs, I've sold about one hundred of these towers to gardeners who pick them up from my front yard. I spread the word online and get commitments via email so I know how many to make in advance. I charge $16 apiece and make about $75/hour over one weekend of light work and friendly garden chitchat.)

The rolls are heavy, so get help loading and unloading them from your pickup truck (though I'm able to drag and drop it from the truck by myself and then I can roll it around on the ground without a helper).

SUPPLIES NEEDED

- 1 or more rolls of remesh
- Bolt cutters—the longer the handles are, the less you'll have to bend over to cut the remesh
- At least 2 cinderblocks or other weights—milk crates or totes weighted down with bricks, etc.—to keep the remesh from rolling back up
- Work gloves
- 4-foot (1.21 m) lengths of rebar to anchor the finished towers
- Tie wire, zip ties, or twine to secure the anchor rebar to the towers

DO-IT-YOURSELF TOMATO TOWERS

The heavy rolls are best transported in a pickup truck, but most suppliers can deliver them for a fee.

1. Lay the roll of mesh flat on the ground. My long concrete driveway works well.
2. Have one helper stand on the end of the mesh (or if working alone, as I often do, put a cinder block or loaded 5-gallon, or 18.92 L, bucket down to keep the wire from rolling back up).
3. Unroll the mesh a distance of 10 or 20 feet (3 to 6 m) and hold that end in place with another cinderblock to keep it from rolling back up.
4. Count a length of 11 full 6-inch (15.24 cm)-wide squares: That's a total length of 5½ feet (1.67 m).
5. Cut the mesh wires with bolt cutters so the last row of your 11 squares has one edge with lots of protruding wire prongs—what I call "fingers"—and the other edge is smooth and fingerless.

Behind John, a bucketful of bricks keeps the remesh from rolling up. A cinderblock next to the roll keeps it from moving too. A bolt cutter is strong enough to cut the sturdy wire to length.

6. The wire is springy and the cut edges are sharp and often rusty, so *carefully* pull up the edges until they meet to form a cylinder.

7. Fold each protruding finger entirely over the opposite smooth side: This creates a tube with an 18-inch (45.72 cm) diameter that stands 5 feet (1.52 m) tall.

8. Viewed from the end, each tower will come out teardrop shaped, so press down lightly where the edges meet until the tower forms a more circular profile.

9. Optional: Cut a third of your towers with 10 squares, another third with 11 squares, and a final third with 12 squares. That way, when it comes time to store the towers, you can nest 3 of them together like Russian nesting dolls. With this option, it's even more important to carefully squeeze down the teardrop shape into a pretty good circle so the smaller towers fit easily into the larger ones.

Often, my indeterminate plants grow taller than 5 feet (1.52 m), but I just let the vines dangle over the sides. The growing tips often reach halfway down the outside of the towers before a frost kills them. Any trellis system that keeps the fruit from touching the ground is a success.

A tomato tower with a smaller diameter can fit over a 5-gallon (3.7 L) bucket and not flop over.

A FINAL TOUCH ON TOP

Some people stake each tower to keep the wind from blowing over the plants. Long tent stakes can work, as can short pieces of rebar. I've even used the cutoff handles of worn-out brooms and other hand tools to keep my tomato towers from flopping over. But I think even that is too much extra work.

We grow four slicing tomato plants in four towers arranged cheek by jowl in a square (growing a set of three touching each other in a triangle would work too). I tie a 3-inch (7.62 cm) piece of jute string at each of the four points where the tops of the towers meet. This creates a larger, more stable footprint that won't blow over. And it takes much less time, trouble, and materials than staking each tower. Which, of course, leaves more time for battering up and eating a plateful of fried green tomatoes.

A trio of bricks can keep the tower from blowing over.

LET'S
do a
PROJECT

Tomato House

KNOWING THAT TOMATO PLANTS EVOLVED IN A MEDITERRANEAN-LIKE CLIMATE WITH DRY SUMMER AIR AND NOT-SOGGY SOIL, IT'S CLEAR WHY THE BRITISH ONLY STARTED ENJOYING TOMATOES IN THE 1800S. THAT'S WHEN THE WEALTHY COULD AFFORD THE BIG SHEETS OF GLASS NECESSARY TO BUILD GREENHOUSES THAT KEPT THE RAIN OFF THEIR TOMATO PLANTS.

Greenhouses are expensive structures so many modern farmers in high-rainfall areas use plastic-covered "hoop houses" to grow these finicky plants. Hoop houses are made of curved hoops of white PVC plastic or metal pipe covered by plastic sheeting to create better conditions:

- The sides are open for ventilation
- The top is rainproof so the leaves don't get wet
- The tomatoes are grown in good soil with a drip line or soaker hose to give the roots the right amount of moisture but not so much that root rot starts

Hoop houses are useful for tomato growing, but their plastic components bring a particular kind of ugly to the home landscape, and the thin plastic sheeting has to be thrown away and replaced every few years—a big investment in time, money, and landfill space.

In tall, agricultural hoop houses and greenhouses, tomato plants are often pruned down to a single stem which is then looped around a length of twine suspended from the ceiling. Chris spent a summer working in a greenhouse like this in Vermont. They had to bring in colonies of bumblebees to help with pollination.

I figured I could come up with something better. Not a greenhouse. Not a hoop house. I call it a tomato house. Like a hoop house, it keeps the leaves dry, keeps the soil just the right amount of moist, and is open on the sides to let breezes in. And, like a greenhouse, it is an attractive garden feature.

But unlike a greenhouse or a hoop house, it is easy to put up, easy to take down, and easy to store out of the weather in the off-season. The rigid panels of polycarbonate that deflect the rain are cheaper than glass and last many times longer than plastic sheeting. Polycarbonate panels are rated to last a decade in the sun, but since they will be stored away for half the year, they should last twenty years.

Here are two kinds of tomato houses. For the first one, a shed roof of polycarbonate panels is supported by a very simple structure of pressure-treated wood. This one is for gardeners without carpentry skills. Anyone who can learn to drive screws with a drill and cut 2 × 2-inch (5 × 5 cm) pieces of wood with any kind of saw—a handsaw, a jigsaw, a circular saw, or even a rented miter saw—can install one of these in a couple of hours.

The finished framing.

SUPPLIES NEEDED

To span a 5 × 5-foot (1.52 × 1.52 m) raised bed with a simple shed roof, you'll need:

- Power drill to drive screws
- Handsaw to cut the lumber pieces to length
- 7 pieces of 2-inch × 2-inch × 8-foot (5 cm × 5 cm × 2.43 m) pressure-treated wood
- One 1-pound (454 g) box of 3-inch (7.62 cm) exterior-grade screws
- One 1-pound (454 g) box of 2-inch (5 cm) exterior-grade screws
- Enough polycarbonate panels to cover the bed

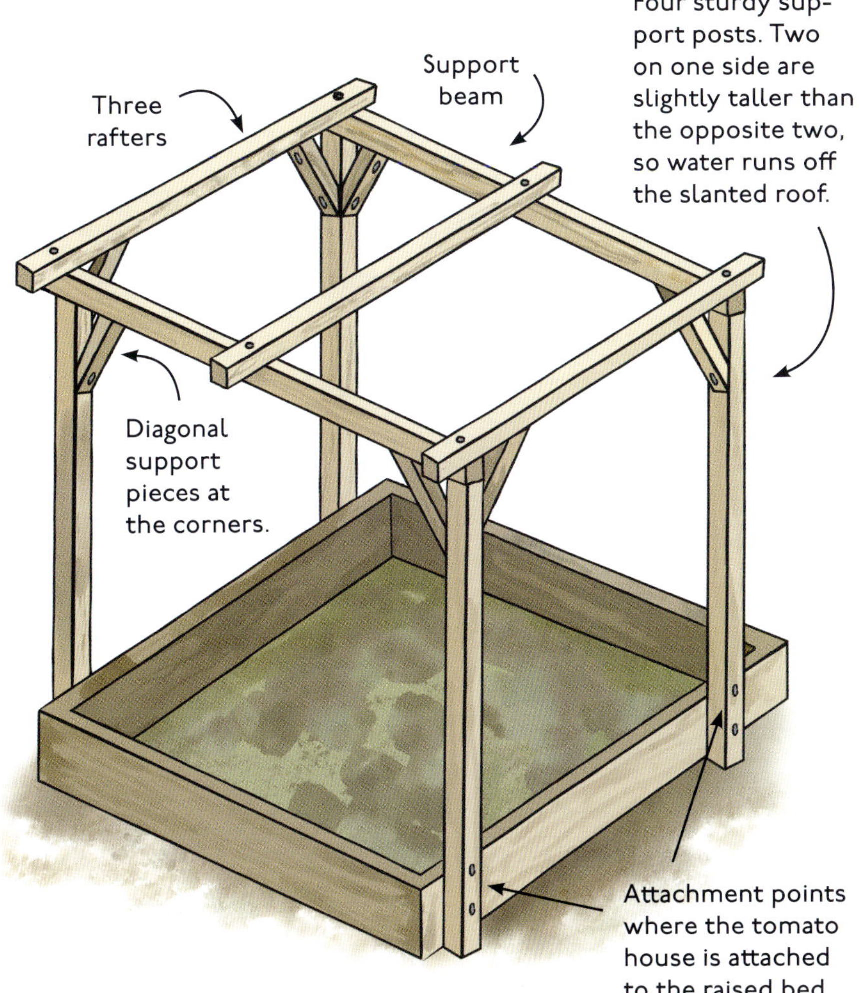

DEFINE THE FOOTPRINT

1. Calculate how big an area of your tomato bed you want to protect from the rain.
2. Buy enough polycarbonate to cover an area at least a few inches wider on all sides.
3. Cut four posts to support the shed roof. Two will need to be 8 to 12 inches (20.32 to 30.48 cm) taller than the other two so the flat—but not level—roof will shed water. Allow enough length for the bottom 8 inches (20.32 cm) of the posts to be screwed to your wooden raised bed.
4. Screw the four 2 × 2s (38 × 38 mm) to the sides of your wooden raised beds—with a pair of 3-inch (7.62 cm) screws in each one to act as posts to support a shed roof that's at least 1 foot (30.48 cm) higher than your tomato plants will be.

Short pieces of diagonal wood at the corners keep this simple tomato house from bowing and twisting.

A pair of screws at the base of each small post hold the tomato house upright against the wooden raised bed.

FRAME THE ROOF

1. Cut a pair of 2 × 2s (38 × 38 mm) long enough to support the width of the polycarbonate roof. We'll call these beams.

2. Screw one beam over the tops of the two tall posts. Ideally, drill a pilot hole in the beam that's slightly wider than the screw threads. Drive one 2-inch (5 cm) screw through each beam-and-post connection.

3. Repeat step 3 over the lower pair of posts.

4. Measure and cut three 2 × 2 (38 × 38 mm) rafters to be the support for the polycarbonate panel(s). Two rafters will rest on the beams, right over the posts. Drill pilot holes in the rafters and install them with 2-inch (5 cm) screws. Install the third rafter at the center point of the two beams. Drill pilot holes in and secure it with 2-inch (5 cm) screws.

5. Cut eight pieces of 2 × 2 (38 × 38 mm) that are 6 to 8 inches (15.24 to 20.32 cm) long. Cut them with a 45-degree angle at each end. These will provide the diagonals that keep a built structure from tilting into a rhomboid shape and leaning, perhaps to the point of falling down. Drill pilot holes in each end of these diagonal pieces—or you could call them struts. Using 2-inch (5 cm) screws, install one at each corner, connecting the post and the beam. Put another one at the same corner, connecting the post and the rafter. When all eight struts are secured, you can work on the roof.

6. If you need to cut the polycarbonate to fit the structure, put it on a couple of sawhorses with a couple pieces of 2 × 4 (38 × 90 mm) to keep it from sagging. Measure and make your mark with a pen and a straight edge. Cut along your mark with an electric jigsaw or a hacksaw. Anything with small teeth will work well.

7. Put all the roof pieces in place to make sure everything fits the way it should. Hold them in place with clamps. The polycarbonate is only 3⁄16-inch (0.47 cm) thick, and it's softer than wood, so you don't need long screws to hold it in place, and you won't need to drill pilot holes. Every 8 inches (20.32 cm) or so secure the roof to the rafters with 1⁄2-inch (1.25 cm) screws.

8. Double-check all your connections to make sure you haven't missed anything. Because the 2 × 2s (38 × 38 mm) are thin, they will have a bit of flexibility, but if all the screws described here are in place, it will stand up to anything short of a hurricane or a tornado.

9. At the end of the growing season, you can back out and save all the screws and stash all the parts in a shed or some other place out of the sun and out of the rain and out of the way, until next tomato season arrives.

LET'S
do a
PROJECT

Advanced Bamboo Tomato House

I'M ALSO INCLUDING INSTRUCTIONS FOR A SECOND TOMATO HOUSE: A PRETTIER ONE THAT TAKES A HIGHER SKILL LEVEL. IT'S MADE OF TIMBER BAMBOO, REBAR, SCREWS, AND POLYCARBONATE. THIS SECOND HOUSE IS MORE CHALLENGING—THAT IS, EXCITING—FOR SOMEONE WITH MODEST CARPENTRY SKILLS, ACCESS TO TIMBER BAMBOO WITH A DIAMETER OF 3 TO 4 INCHES (7.62 TO 10.16 CM), AND A SECOND SET OF HANDS TO ATTACH THE GABLE ROOF.

I don't get into as much detail on the construction steps for this one because that would form a small book itself. If you're not familiar with carpentry, some of it may not make sense. Nonetheless, from the supplies list, photos, captions, and modest directions, a decent carpenter—or an ambitious novice—will be able to build this beautiful and store-able temporary structure.

SUPPLIES NEEDED

To cover a 5 × 5-foot (1.52 × 1.52 m) bed you will need:

- Power drill
- 3-inch (7.62 cm) hole saw for the power drill to make the curved cuts at the end of the bamboo
- One 1-pound (454 g) box of 2-inch (5 cm) exterior-grade screws
- 4 pieces of steel rebar, ⅜ inch (0.95 cm) thick (¾ inch, or 1.9 cm, thick is fine too) and 4 feet (1.21 m) long
- Two 3-foot × 8-foot × 6 mm (91.44 cm × 2.43 m × 6 mm) polycarbonate panels made for exterior use (be sure to keep track of which side is resistant to the UV radiation from the sun that damages any kind of plastic)
- 6 pieces of timber bamboo, about 10 feet (3 m) long and 3 to 4 inches (7.62 to 10.16 cm) in diameter

LAYOUT THE FOOTPRINT

1. Determine the corners of your bamboo tomato house.

2. In the spot for each corner post, drive a 4-foot (1.21 m) piece of rebar, vertically, into the ground 12 to 18 inches (30.48 to 45.72 cm) deep. There will be 30 to 36 inches (76.2 to 91.44 cm) of rebar above ground.

3. Cut four pieces of timber bamboo long enough to be posts that hold up the lowest part of the roof a bit higher than the tomatoes will grow within it.

4. Use the 3-inch (7.62 cm) hole saw, mounted on a drill, to cut off the top of each post. You may have to cut partway from one side, back out the saw, and finish the cut from the other side by lining up the drill bit with the hole from the first cut. The two cuts may not be perfectly aligned, but no one will be looking closely enough to be bothered by that. As long as the top of each post has a semicircular cutout to support the two horizontal bamboo beams that go over the tops of the posts.

5. Hold the first bamboo post vertically over the rebar. Drop it or force it down onto the rebar—like a sleeve over a skinny forearm—and let the rebar punch its way through the first of the horizontal partitions inside the bamboo. Keeping the bamboo post as vertical as possible, drop it or force it down over the rest of the rebar, through a total of two to four partitions, until the bottom of the bamboo touches the ground. The holes punched through the partitions will be small enough to hold up the post pretty close to vertical.

6. Double-check your measurements and repeat this with the other three bamboo posts. Make sure to start with the semicircular cutout on top of each post oriented to support a horizontal beam.

7. Measure, mark, and cut two beams with diagonal cuts of about 45 degrees. This can be done with a handsaw (or a miter saw if you have one). You may want to practice with some scrap pieces to get the angle right at both ends.

8. Lay the two beams on all four posts and check for level. You may need to put some kind of shim under one or more of the posts to get them the same-ish height. When they look okay, use 2-inch (5 cm) screws to secure the beams to the posts.

9. Say your beams are running north-south. Now, you need two more beams running east-west. Use the hole saw to cut both ends of the two east-west beams to fit between the posts, lower than the north-south beams. You'll need clamps or a second set of hands to hold these beams until you can tie them into the posts with 2-inch (5 cm) screws.

Short diagonal pieces of bamboo keep the horizontal beams and vertical posts square.

10. Cut eight bamboo struts/diagonals with 45-degree ends 8 to 16 inches (20.32 to 40.64 cm) long from the long point to the long point. Drill pilot holes and install two at each corner with a 2-inch (5 cm) screw at each end. Check each corner to make sure it's reasonably close to being square. A helper would be valuable here.

11. You're not done yet, but take a look at your work and pat yourself on the back. That wasn't easy, especially when it's your first time. But at this stage, you're seeing a smaller version of the first structure I built using this technique: the secular/Hoopa/sacred space I made for our outdoor wedding on a friend's farm. Instead of being a rectangle, it had a trapezoid footprint—like a theater stage—with bright orange fabric hanging down both sides, like stage right and stage left—with the wedding party in the center as the main characters in a performance. Which is what a wedding is, when you get right down to it.

12. With a small-tooth saw, cut the polycarbonate to size to make a gabled roof with modest eaves.

13. Cut a 2 × 2 (38 × 38 mm) to be the ridge beam—turned on a diagonal—and with diagonal 45-degree ends. With a helper holding the ridge beam to the edges of the roof panels, drive one screw on each side about every foot (30 cm) or so. Don't drive them all the way in. You want some flex in the roof until you've also connected it to your two horizontal beams.

This image shows the drip line on the coffee chaff mulch beneath the bamboo tomato house. The chaff below the drip line is wet. The chaff above the line is dry. But the roots of the dwarf and plum tomatoes all get plenty of moisture from the soaker hose.

14. With a helper under the center of the tomato house on a short ladder holding up the ridge beam, direct them so the ridge beam is level-ish and centered between the two beams. Mark where the beams meet the roof panels.

15. Cut a pair of 1 × 2s (19 × 38 mm) to match the length of the roof panels. Attach them with screws to the roof panels at your marks.

16. With your helper holding things together—with or without assistance from clamps—screw the 1 × 2s (19 × 38 mm) to the horizontal bamboo beams.

17. Last step: With the roof panels secured to the beams, get on a ladder outside the tomato house and gently tighten the screws holding the roof panels to the ridge beam. Get off the ladder and put your tools away. Then, take a selfie with the tomato house behind you and broadcast it on the internet.

Dyson spreads coffee chaff mulch around the finished bamboo tomato house (bottom corner) and the simple wooden tomato house (upper corner). The top left corner of the garden shows indeterminate tomato plants growing over the top of their tomato towers. The center right corner of the garden shows a late planting of indeterminate plants just getting started.

The completed timber bamboo tomato house keeps the rain off these paste tomatoes in short square tomato towers. Removing about three dozen screws makes the bamboo tomato house store-able in the off-season.

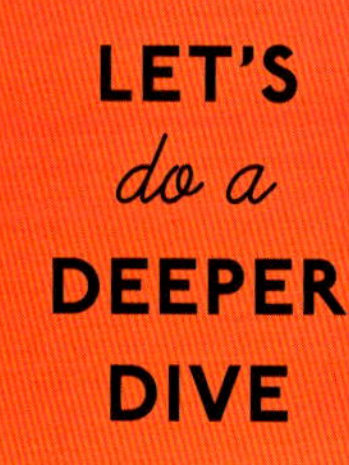

Are you a complete novice in carpentry? To bump up your carpentry skills and make building either of the tomato houses easier to manage, I recommend the chapter on carpentry for beginners in my book *Hentopia*. I cover the exact tools and techniques that will keep you safe and out of trouble. Search for "Hentopia" on YouTube to learn more about the book.

CHAPTER 6

Tomato Pests, Diseases, and Challenges: What They Are and How to Beat Them

IT'S NOT JUST PEOPLE WHO LOVE TO EAT TOMATOES. DISEASES AND BUGS LOVE TO EAT THEM TOO. BUT I HAVE TWO BITS OF GOOD NEWS.

First, there are many simple steps before and during the growing season to prevent diseases.

Second, even if your prevention techniques are less than perfect, tomato plants are tough. As long as you have made some effort to diminish the impact of bugs and diseases, they will produce a good number of fruits, even if the plants don't look perfect.

Rather than doing a painfully long rogue's gallery of every pest and disease, I'm going to break them down into five general categories:

1. Leaf diseases
2. Root diseases
3. Fruit problems (the fruits rarely suffer from actual diseases)
4. Bad bugs
5. Good bugs

The most common problems occur on the leaves, and the most familiar example is a fungus called early blight. It looks like one or more very irregular brown circles on the leaves with visible concentric rings. The rings are like growth rings on a tree. And as on a tree, new rings are added over time. If unchecked, each rain—or sprinkler dance—will bounce the fungal spores from the soil up onto the plant's lower leaves. And the next shower will then bounce spores from the lower leaves onto the mid-level leaves, and on up the ladder of fungal success until there is a steep decline in the health of your tomato plant. Diminishing early blight, and many other tomato diseases, is a major key to success.

Preventing Leaf Diseases

TIP 1

Some tomato varieties tout being resistant to various diseases. This information is worth paying attention to when you select your varieties and if you know you've had a problem in past seasons with specific diseases—but "being resistant" doesn't mean "immune." It just means the disease will have a harder time overwhelming your plants. It's a good first step.

TIP

Add a couple inches of compost to your vegetable beds at least once a year. Compost adds little in the way of nutrients compared to fertilizer, but it adds beneficial microbes that can outcompete diseases and keep their numbers down. It's similar to a person taking a probiotic full of beneficial microbes to outcompete bad microbes in their gut. Both practices help build an immune system. Compost also supplies food to earthworms and other soil organisms that aerate the soil, providing more oxygen to plant roots, which also helps plants build up their immune system.

TIP 3

Don't add too much nitrogen. Too much nitrogen for plants has an effect similar to too much sugar for kids. Too much sugar will certainly make kids grow—in a bad way. And it will certainly get them energized—until they crash. Too much sugar is clearly a net loss for kids. Ditto regarding too much nitrogen for plants. They will become a beautiful, dark green. They will grow more quickly. But their immune system will be compromised. And that dark green growth is seen as attractive by diseases and bad bugs. Don't deprive your plants of nitrogen fertilizer, but do use an organic fertilizer that becomes available over the course of the growing season rather than all at once, like synthetic salt fertilizers.

Regardless of how much compost or nitrogen you apply, there will likely be some number of disease microbes in your soil. Even if you grow tomatoes on that soil for the first time, many diseases come in on your transplants as the greenhouses they were started in likely have early blight and other diseases. But you can, literally, provide some shielding for your tomatoes. ***Immediately after setting your transplants, apply a 2-inch (5 cm) layer of organic mulch—shredded leaves, coffee chaff, shredded pine bark, or 5 inches (12.7 cm) of wheat straw.*** That way, every time it rains, the organic mulch will absorb the force of the raindrops. Without the rain hitting the soil directly, fungal spores won't be able to bounce up onto the lower leaves and climb their way up the ladder of pestilent success.

Of course, there will be times when rain is scarce and you want to irrigate. So, ***before you put down that layer of mulch, lay out a soaker hose*** (more about them in chapter 7). That way, when your plants aren't getting enough rain, you can turn on your faucet and apply enough water to maintain steady growth of the plants and their fruit. With that layer of mulch covering the soil and the soaker hose, you'll not only keep fungal spores from getting on the leaves but also you'll keep the moisture in the soil from drying out the roots.

TIP
6

So far, you've enriched the soil, employed a tidy soaker hose instead of a sloppy sprinkler, and shielded your plants from fungal spores with a layer of organic mulch. One other easy way to reduce the degree of what horticulturists call "disease pressure" is to ***remove the lower leaves from your tomato plants***. That way, if any spores do get past the mulch shield and get splashed by the rain, the lowest leaves will be too high for the backsplash to reach. Unfortunately, these techniques rarely give you 100 percent control. So, when you cut off the lowest leaves, don't use your fingers or pruners. There may be spores on those leaves that you can't see yet, and you don't want to spread them around inadvertently. My practice is to use a sharp steak knife on each plant that I can then run through the dishwasher. Steak knives can be purchased cheaply at thrift shops.

Preventing Root Diseases

Every step in the previous section about preventing leaf diseases also applies to combating root diseases—except the pruning of lower leaves.

The indication that your plant has root rot is that it will start to wilt as if it hasn't had enough water because that is exactly what is happening. A fungus in the soil has overwhelmed that tomato plant's immune system and is killing the roots bit by bit. ***If the plant doesn't respond to watering by the next morning, dig it out of the garden and put it in a trash cart that goes to the landfill, or burn it.*** You don't want to compost these plants or put them in your yard waste bin as doing so will just spread the diseases. You may want to be proactive and also remove the neighboring tomato plants. They may also be infected but not showing symptoms yet. Once removed, add a good load of compost to the soil. Those good microbes will compete with the fatal fungi.

Blossom end rot (BER) occurs early in the season and is caused by insufficient water that's required to distribute calcium to the fruit. Boosting irrigation is the answer.

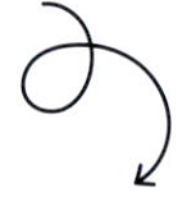
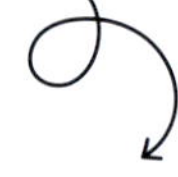

Preventing Fruit Ailments

I didn't use "diseases" in this heading. It's possible to have fruit diseases, but it's not very common. *But you will likely see problems on your fruit that are not disease related.* Here's how to handle them.

BLOSSOM END ROT (BER)

The most distressing and common fruit problem is called blossom end rot (BER). If this appears, symptoms will show on some of your earliest fruits and will often abate as the season progresses. BER appears as a black patch covering the bottom of the fruit (the end that the flower blossom was originally attached to, hence the name). It is not a disease. Sometimes, the stem end will ripen and be good enough to eat after cutting off the blossom end. But when I see it, I just pop off the whole tomato and feed it to the hens.

BER shows up among the first fruits when the plant is younger/smaller and can't get enough water to draw calcium from the soil at the same rate that the fruits are growing. With enough rain or irrigation, the plant will soon outgrow this problem. Technically, this black spot is caused by a shortage of calcium in the developing fruit. But adding lime to the soil or spraying liquid calcium on the plant won't help this season. The real problem is that the plant didn't have access to enough water to carry the calcium into and through the plant. How to avoid that? One inch (2.54 cm) of rain per week in spring and fall is plenty for most vegetable gardens. That provides just over 50 gallons (189.27 L) of water per 100 square feet (9.29 sq. m) per week. Enough for steady growth.

In a hot summer, about 2 inches (5 cm) of rain or irrigation per week provides just over 100 gallons (378.54 L) per 100 square feet (9.29 sq. m). Quick math: 1 gallon (3.78 L) per square foot (0.09 sq. m) *in summer* for vegetables. The best response is to put a timer on your spigot that applies enough water: ½ gallon (1.89 L) per square foot (0.09 sq. m) per week in spring and fall. In summer, double that.

Another way to picture this:

a A single 5 × 5-foot (1.52 × 1.52 m) bed = 25 square feet (2.32 sq. m) and needs about two and a half 5-gallon (18.92 L) buckets of water per week in spring and fall.

b The same 25-square-foot (2.32 sq. m) bed in summer needs about five 5-gallon (18.92 L) buckets of water per week. You can see why applying water twice a week makes more sense.

CRACKING FRUIT

Crack lines in the shoulder of your fruit is, again, not a disease but an indication of uneven watering. With very little water, a tomato fruit grows slowly. Follow that with a lot of rain and the fruit will absorb too much moisture and grow more quickly than the fruit's skin can accommodate. Hence, it cracks. Steady watering, twice a week from rainfall and/or irrigation, will minimize or eliminate cracking.

SUNSCALD

Sunscald on the shoulders of fruit is not a disease. Too much sun damages the pigment and the cell structure, making the shoulders of your tomatoes pale and discolored. When I see that, I pop the affected tomato off the vine and toss it to the hens.

The tomato on the upper left has what's called "cat-facing." The marks don't harm the fruit or the flavor. They are merely injuries, perhaps from insect bites, when the fruit was small and green. It's just a cosmetic issue.

Tomatoes—*They Aren't* for the Birds

Birds often enjoy feasting on tomatoes as they ripen. You may be tempted to cover your plants with bird netting, but one of the dirty little secrets of bird netting is that it can work too well. If you've visited a netted blueberry hedge and found a dehydrated, dead bird with its feet caught in the netting, you know what I mean. Tightly securing the edges of the netting helps, but it's lots of work and impedes harvesting.

SAFE SNAKES

I prefer scaring away the birds with inflatable snakes. This saves about 95 percent of the fruits. Plastic snakes from museum gift shops, or fake snakes from garden centers that blow up like a colorful, skinny, curvy beach ball, are easy to hang over a tomato plant or trellis. We don't put them out until just before tomatoes get to be full size. To sustain the ruse, we move them *every other day*, a frequency that lines up nicely with our picking schedule.

Another way to deter birds: I drive in a few pieces of rebar that are taller than the tomatoes, then tie off and twist some mylar ribbons over them. The flashing light from the twisted ribbons reminds birds of light flashing off the eyes of predators such as hawks.

Is the Word "Deer" *Just* a Euphemism for "Long-Legged Rats?"

Deer are common in so many regions now. Protecting your tomato garden from them is essential. Here are a few of my favorite methods.

DEFENSE WITH NO FENCE: DOGS

At an organic farming conference a few years back, I spoke briefly with a farmer who kept deer off his land with his pet Chihuahuas. Lots of them. I recall that he had maybe a dozen Chihuahuas posted along his property line, four per doghouse surrounded by a chain-link fence. Whenever deer came within sight or smell of the tiny dogs, they let out an anxious string of barks. But that isn't what stopped the deer. When the farmer or his wife heard the barking, they opened the backdoor of their house and let another dozen of their favorite Chihuahuas out to chase the deer away.

If there are very few deer in your area, just the smell of a dog may be enough of a deterrent, but once their numbers get higher, they'll figure out that your dog is indoors or with you on an errand and they'll know that a doggy smell is no real threat. So, keeping one or more outdoor dogs may be necessary.

DEFENSE WITH NO FENCE: REPELLENTS

I travel around the country giving garden programs on dealing with deer in the garden, and I always ask what the local experience is with repellents. Invariably, different gardeners toss out the names of different repellents, but the consensus is that, after a year or two, the deer might develop tolerance for one brand so you'll need to switch brands over time.

DEFENSE WITH NO FENCE: HEDGES

Almost any vegetable or fruit that humans like to eat is going to be appetizing to deer. And deer are skinny and wily enough to slink through most hedges when they smell some tasty plants. But, if you have modest deer populations, a thick hedge of plants that deer really don't like may deter them from visiting small gardens. At one of my garden programs, I met a gardener who keeps a tight hedge of lavender around her roses. The strong flavors and fragrance of lavender are turnoffs for deer—and her roses thrive. The bottom line in most cases of deer predation is that their numbers are rising and they are hungry. Perhaps the most enlightening part of my program (or at least it gets the most laughs) is the back page of my handout: It contains recipes for venison. I think having more people harvesting deer and cooking up some venison (hey, it's fresh, local, and organic) is the long-term answer to the problem of deer invading the farm or garden. But until that day, you may need a fence.

I'll confess I'm looking for a client to hire me to build a bespoke roofed deer stand on top of the house, with room for a chair, a cupholder, and a place to store a crossbow.

A deer could jump over a fence this low, but they avoid this garden because the raised beds create two hazards that deer want to avoid: 1) hopping onto the raised beds would be a trip hazard and 2) the narrow space won't allow them room for a running start to escape.

DEFENSE WITH A FENCE

Sometimes, deer repellents and deer-resistant plants just aren't enough and a fence is called for. However, deer fences don't have to make your yard look like a prison farm.

Many city, suburban, and rural gardeners have installed deer fences that are either nearly invisible or very attractive.

You can enclose part of your garden—say, your prize roses—or your entire property using these options:

- A 7-foot (2.13 m)-tall black plastic fence that is almost invisible. Deer can jump higher than 7 feet (2.13 m) but because their eyes lack the fine detail that human eyes can pick up, they aren't sure just how high the fence is so they raid the neighbor's yard instead.
- A 6-foot (1.82 m)-high solid-wood privacy fence will deter deer because they can't see what's on the other side. They don't want to jump into a sketchy situation.
- Deer can jump a 6-foot (1.82 m)-high wire fence, but if the enclosed garden is small enough—say 12 × 12 feet (3.65 × 3.65 m) or less—the deer will realize that, even though they can get in, they won't have the room to get a running start to jump out.
- Even a 4-foot (1.21 m)-high picket fence will keep deer out if there's a matching fence set back 6 feet (1.82 m) away, in parallel. This creates too wide a distance for the deer to jump. The long, narrow space between the two fences can also serve as a dog run, since dogs and gardens don't always mix well either.

THE SIX STEPS OF IPM

Follow these six steps every time you discover a pest in your garden and aren't sure what to do about it:

1. DETERMINE ACCEPTABLE PEST LEVELS.

How many bad bugs is too many? You want to take action only when your crop could be overwhelmed by the number of pests present. So, how often do I spray or do anything for bad bugs? I haven't sprayed in I don't know how long. When I see a caterpillar eating a hole in a tomato, I toss that tomato to the chickens or into the compost. I don't spray. When caterpillars are eating cool-season plants, like broccoli or collards, I don't spray. I ignore them because they mature and fly away or are parasitized or are eaten by birds. The main thing is that most plants outgrow the damage if they are growing in soil that is highly organic and not too high in nitrogen. You want to allow time for nature—parasites, birds, your fingers—to redress the balance. If it doesn't look like the plants will die, don't spray.

2. EMPLOY PREVENTIVE CULTURAL PRACTICES.

Talk to other local gardeners so you can choose varieties to grow that do well locally. Don't plant varieties that have performed poorly for others. Make sure there is plenty of organic matter in the soil to feed the earthworms and microbes that create optimal conditions for plant roots. Use an appropriate amount of nutrients so your plant has its own excellent immune system. Mulch properly, clean your clippers between uses, and use other preventive measures to prevent issues in the first place.

3. CONDUCT REGULAR MONITORING.

Observe your plants every two or three days. You don't want to be caught by surprise. You also don't want to drive yourself crazy checking them every day. It will look like they aren't growing. Checking them twice a week lets you stay ahead of real problems and doesn't overwhelm you with worry.

4. USE MECHANICAL CONTROLS.

Handpick bugs if they're big enough, like caterpillars. Hand pull weeds when they're small and put them in a pot to feed to your hens. Snap off leaves that have diseases so the rain won't spread the spores. Use row covers to protect pest-prone crops. Mulch to suppress weed seeds by denying them sunlight.

5. UTILIZE BIOLOGICAL CONTROLS.

Let the good bugs (a.k.a. beneficial insects or allies) have first crack at your bad bugs. They need to have enough bad bugs around as a food source to keep their populations stable. If plants are in decline, use biological pesticides rather than chemical ones. Bt (*Bacillus thuringiensis*) is an example of a biological control product. It's made from a bacteria that kills specific bad bugs, like caterpillars and mosquito larvae in water gardens. Another example is milky spore (*Paenibacillus popilliae*), which is a bacterial product that will kill Japanese beetles in their underground larvae stage. Biological controls will not harm other living creatures, only the targeted pest.

6. RESPONSIBLE PRODUCT USE.

I won't be talking about using synthetic chemical sprays here, but this final step is really about not going overboard, regardless of what technique or material you are using . . . even if that product or material is organic.

A serving tray with a mesh cover keeps houseflies, fruit flies, and other household pests from spoiling tomatoes on the counter. A towel under the fruits captures any liquid if one goes bad prematurely.

Beneficial assassin bugs in their smaller, bright red larval stage. They extend the lance from under their chin to suck juices out of aphids and other soft-bodied pest insects.

HOW TO GET A HANDLE ON CATERPILLAR PESTS

Caterpillars are among the most common insect pests of tomato plants. If sections of a leaf are missing, it's not a disease. It's a sign that a moth has laid eggs that have hatched into caterpillars, though many gardeners call them "worms." These worms—tomato worms, hornworms, tomato fruit worms, etc.—have a voracious appetite. The first, and best, control is scouting your plants regularly and handpicking the worms. You can feed them to your hens (they're carnivores!) or squish them and donate them to your compost bin, or put them on nongarden plants that are in the nightshade family, such as horse nettles, so they can grow up to be night-flying pollinators (called hawk moths, in the case of hornworms). If you're too squeamish for that, apply the organic pesticide Bt either in powder form or liquid form. It's a bacteria that affects only caterpillars and won't hurt anything else. Spritz or powder it under the leaves (so it's less likely to disappear from rain) and over the fruits. When the caterpillar eats the Bt it gets sick and dies in a few days.

Additional Caterpillar Tactics (Especially for Novices)

- Identifying the species of caterpillar feeding on your tomato plant is interesting, but not necessary. They may be eating the fruits or the leaves. Sometimes, you'll see little brown turds they've left behind. Smaller caterpillars may spin a small, webby tent to hide under while they eat.

- If you see a caterpillar of any size in the tomato garden—on a fruit or on a leaf—feel justified in squishing it before it can do more damage. Tomato plants are not a host plant for any butterfly caterpillars. If you're squeamish, squeeze the caterpillar between a leaf or inside its tent. Drop it on the ground to decompose.

- If you find a round hole in a tomato, it's the work of a caterpillar and the fruit is spoiled. It may still be inside eating, or it may have moved on to another fruit or to pupate into an adult moth. Toss those damaged fruits. Tomato seeds in compost may still be viable and you're going to get lots of tomato seedlings that may not come true. I give my caterpillar-eaten fruits to the chickens. Their gut will dissolve any seeds and they will devour any caterpillars still inside.

    - The exception to this rule is that if you find a caterpillar with a number of tiny, silky white cocoons about the size of a grain of rice on its back, let it live. The tiny cocoons are beneficial insects at work! A tiny adult wasp—that doesn't sting people—has inserted its eggs into the caterpillar. The hatching larvae eat the caterpillar's internal organs, then crawl out to create their cocoons. Let them pupate so your garden will have more caterpillar killers.

In the home garden, caterpillars can be easily managed with the WALA WALA (walk around, look around) technique and smooshing them.

HOW TO GET A HANDLE ON THE MANY OTHER TOMATO PROBLEMS THAT AREN'T CATERPILLARS

you see leaves that are mottled, blotchy, distorted, faded, speckled, or discolored:

look closely for insects that are eating or sucking juices from those leaves and, perhaps, spreading disease. They could be whiteflies, aphids, spider mites, cucumber beetles, and other insects smaller than your pinky nail.

- **Option 1:** If you see a second type of insect—one or more of your beneficial insect allies, like ladybugs or assassin bugs, leave everything the way it is. Let your allies feast and multiply so you'll have less pest damage next year. You'll have even more allies—and they'll often appear earlier in the season—next year.
- **Option 2:** If the pests are only on a small number of leaves and have done little damage, leave them until your next WALA WALA and, perhaps, they will attract some allies in the meantime. This will mean you can happily do nothing and let the good bugs find them in a few days.
- **Option 3:** If the damage is extensive and you don't see any allies, feel free to strip all the damaged leaves with their pesty passengers, smoosh them, and/or just toss them in the compost.
- **Option 4:** Brush all the pests into a jar of soapy water. The soap will block their breathing pores and kill them.
- **Option 5:** Spray the pests in the morning with a store-bought organic soap-based insecticide that will either suffocate them or dissolve their waxy coating and cause them to dehydrate if the day is sunny.
- **Option 6:** You will ruin your chance at gaining allies, but you can spray an organic broad-spectrum herbicide like those based on spinosad. It will kill pests *and* beneficial insect allies so consider this only as your last option.

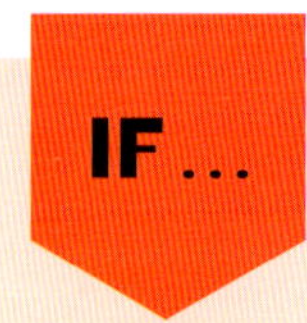

you see lower leaves that are yellowing and/or have brown, gray, black, or yellow spots:

you're probably dealing with diseases like early blight, late blight, septoria leaf spot, bacterial speck, bacterial spot, etc. These are pathogens in soil that can be splashed up onto the lower leaves by rain or sprinklers if the soil isn't covered by mulch. Remove the damaged leaves all the way back to the main stem and make sure the soil is completely covered by a 2-inch (5 cm) layer of natural mulch or a layer of plastic mulch. Also, these problems can be reduced by using soaker hoses instead of sprinklers and by growing tomatoes so they aren't exposed to rainfall by using a greenhouse, hoop house, or my Tomato House (see p. 96). Also, some of these microbes can blow in on the wind, but are often brought in on the potting soil of store-bought tomato transplants. Starting your own seedlings and planting them in virgin soil can keep these microbes at bay for a few seasons.

despite adequate irrigation, the entire plant or a large portion is wilting, even in the morning:

your plant is suffering from pathogens that attack the roots, such as bacterial wilt, verticillium wilt, fusarium wilt, and others. Even tiny worms in the soil called root knot nematodes can cause these symptoms. (Note: On a hot day, any plant—and even a gardener—can wilt in the afternoon sun but then recover as the temperature drops and look fresh in the morning.) The short-term remedy is to yank out all wilting plants and drop them in the trash—not the compost—so those microbes have less of an opportunity to multiply. The long-term solution is to boost your commitment to adding compost to the garden soil. A higher number of species of compost-based microbe allies may outcompete the pest microbes in your garden soil. Another option is to choose hybrid plant varieties labelled as resistant to various wilts.

the fruits show damage like a sunken black spot on the bottom:

this is a sign of blossom end rot (BER—see p. 110). As we've discussed, it's really a sign of too little irrigation early in the season. The plant doesn't have enough moisture to send calcium all the way through the rapidly growing fruits. It's too late to add calcium to the plant or the soil. Just bump up the irrigation. Toss the blackened fruits to your hens rather than the compost.

the ripening fruits have cracks across their shoulders or around their waist:

they are absorbing too much water and what you see is called "cracking." If the weather report calls for a heavy rainstorm, pick your near-ripe fruits and let them ripen indoors and avoid getting bloated and busted. It won't hurt their flavor. The tomatoes we buy in the store that are picked before they're ripe have poor taste and texture because those varieties were developed for surviving bouncing in a truck rather than for good flavor and texture.

your fruits have pale, bleached-out areas:

they have "sunscald" due to not enough leaf cover to shade the fruit from the sun. This can happen when the gardener removes suckers and exposes fruit clusters to excessive sunlight. Studies show that removing suckers isn't necessary. It merely reduces the amount of fruit set and reduces the amount of shade on those fruits, which could lead to sunscald.

your fruits have small white spots:

stink bugs have found your garden. They are bigger than your pinky nail, brown to gray, shield-shaped, and they suck juice from the fruits, leaving nasty white spots. They stink if you crush them—which probably correlates with their lack of predators—and they're hard to kill with insecticides. Carry a jar of soapy water and drop them in or sweep them into the jar to kill them. Then, dump the soapy water and all into the compost.

the ripe fruits have stabby, torn holes:

thirsty birds are drinking the moisture in the fruits. Invest in a charming bird bath—or a fantastic water garden!—which they would prefer.

A tomato worm that got ahead of me. The undamaged parts of the tomato are still edible. Or the whole thing can be tossed into the hen pen. Chickens' favorite food are bugs. Second favorite is fruit.

A roll of burlap secured to stakes will keep rabbits, ground hogs, and similar vermin out of any vegetable garden, including those with tomatoes.

whole fruits, chunks of fruits, and big swaths of leaves are missing around knee height-ish or lower:

one or more species of wild, four-legged mammals are using your yard as a drive-through restaurant. It's likely squirrels, rabbits, possums, raccoons, or groundhogs. I have had luck deterring these vermin—and birds—with plastic snakes strewn around the garden. I've also used colorful mylar tape as a deterrent. Use tall stakes or rebar to run two or more strands of the tape over the top of the garden, above head height, so you can walk under it. Tie the tape to the first stake, then twist it a bit—like crepe paper at the school dance—then tie it off to another stake. As the reflective tape twists in the wind, many critters will interpret the flashing as the reflection of light from a raptor's eye. I also know of gardeners who've deterred these critters with store-bought spray repellents. A simple fence that's 2 to 3 feet (60.96 to 91.44 cm)—low enough for the gardener to step over—can deter rabbits. The others may need to be caught with a live trap and relocated, if that's a legal option where you live.

whole fruits, chunks of fruits, and big swaths of leaves are missing around knee height-ish or higher:

... THEN

you're dealing with deer. Some gardeners have luck scaring them off with sprinklers connected to motion detectors. But some deer figure it out and work around the detectors and/or the spray (which you don't want to wet the tomato plants). Deer can jump higher than 7 feet (2.13 m), but a black plastic mesh fence that high will keep them out because their eyesight isn't sharp the way ours is. They will look up and not know how high the fence is because the black mesh doesn't reflect enough light; they will worry it's higher than they can jump, and they'll get caught in it. If your entire vegetable garden is small—say, 8 × 12 feet (2.43 × 3.65 m) or so—a 6-foot (1.82 m)-high fence will keep them out as they won't have enough running room to jump out. See page 112 for more tips on dealing with deer.

All the Other Insects of Tomatoes

Depending on your region, there may be five or ten other pests, aside from caterpillars, that damage tomatoes, but it's not necessary to learn them all by name. Only a few of them will be in your neighborhood. Their injury to leaves will be somewhat similar and may be described as mottling, blotchy, distorted, faded, speckled, and discolored.

These looks will catch your eye on your walk-arounds. Check the leaves more closely to see if there's one or more—probably more—insects on the damaged leaves. These are your pests. Smoosh them or drop them in a jar of soapy water to suffocate them.

If you're lucky, you'll actually see two different types of insects: a good number of pests, such as aphids or whiteflies, and a smaller number of allies, such as assassin bugs or ladybugs. If you see allies in place, *do nothing*. With enough pests to eat, your allies will stick around, lay lots of eggs, and show up ready and able to work next season. I repeat, if you see allies and pests together, do nothing.

If there are no allies in sight, smoosh as many of the pests as you can scout. If you're feeling squeamish, this is a good opportunity to grow as a gardener. Smooshing bugs has several virtues. It lets you get in touch with your inner ten-year-old, something you will never regret. (I also get in touch with my inner ten-year-old by riding my bike, climbing trees, skipping stones, and fording creeks barefooted. So, smooshing bugs isn't the only option for this worthy goal.) Another virtue of smooshing is you don't have to mess around with buying some spray, filling up a sprayer, shooting said product on all the leaves, whether they have pests or not, and spraying again after it rains. Spraying is not gardening. Smooshing bad bugs is gardening. Also, if you're spraying something—even something organic, like spinosad or insecticidal soap—you're very likely to kill some or all of your allies in the garden stalking your pests. *Sigh*. Spraying isn't gardening.

Good Bugs: Insects as Allies, Not Just Enemies

An adult praying mantis has a huge appetite and will look you in the eye.

As you know, not all bugs are bad, and I've already touched on how helpful beneficial species can be in controlling pests. Let's take a deeper look at the subject when it comes to growing great tomatoes.

Most bugs get a bad rap. Admit it. But as long as they aren't eating you out of house and garden, some of them have a certain charm:

- Ladybugs that land on your sleeve
- Praying mantises that look you in the eye
- Slender lacewings on the other side of the window

- Dragonflies doing their imitation of a helicopter
- Butterflies rewarding us with their beauty for having the good sense to get outdoors
- Those brown-and-black wooly worms in their two-tone fur coats children use to predict the length of winter (fun fact: that wooly worm is a caterpillar that becomes the Isabella tiger moth [*Pyrrharctia isabella*])

Juvenile praying mantises will often let you gently handle them. And sometimes they do a little rocking dance, back and forth.

I invite you to take a more positive view of insects, generally, and to think of them as allies. Even the baddies—in many cases—don't cause enough damage to be worth the trouble to exterminate them.

You might guess I have a soft spot for insects—unless they join the ranks of vermin that eat my plants, then they might find their own soft spot under my thumb.

I had to take an entomology class to get my horticulture degree. I came close to switching my degree: "You mean I get a grade for going outdoors to catch and label insects?!" But the greater lure of garden-making kept me on track.

I think of insects as little dinosaurs. Many are beneficial to the garden. A few are detrimental. But all play a role in keeping things on planet Earth from coming to a complete stop by pollinating flowers, eating other insects, turning dead plants into rich soil, and more.

Of course, local conditions will determine just how bad the baddies are. So, make the most of chances to chat with other, more-experienced gardeners. My experience is that the longer someone has been gardening, the less anxious they get about bugs. Because some aren't as threatening as they seem at first, and because the gradual improvement of your soil gives your plants a stronger immune system to fend off any plant-munching predators, most bugs are not worth worrying about. Plus, those gradual garden improvements also often boost populations of the allies that devour pests, like ladybugs, lacewings, assassin bugs, and others.

A stinkbug contains nasty chemicals that deter predators. Hence its smell when you crush one. And you will have to catch them and crush them yourself, so decide to not mind the smell. Weekly spraying with spinosad will also deter them, but that will also kill some beneficial insects. It's best to smoosh them or drown them in soapy water.

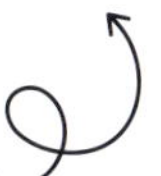

SOME STEPS FOR GARDENING FOR ALLIES

What can a gardener do to have more beneficial insects in the garden?

First thing, consider coming to think of insects in a more positive light, perhaps by reading a book called *Tiny Toilers and Their Works* from 1921 by G. Glenwood Clark, or other books that describe the insects that grow and harvest edible mushrooms that enrich the soil, the African termites that build castles of hardened sand that can support the weight of a man and provide food for aardvarks, the insects that turn scrapings of wood into paper nests while pollinating flowers, etc. Insects are cool.

Think twice before disturbing a spider's web in your garden. They eat lots of insects.

Spraying isn't gardening. Don't give in to the temptation to spray the heck out of every bug you see. That would be as smart as pulling every plant from a bed without knowing if it was a weed or a valuable garden plant. Whether ally or pest, they are all different enough to recognize them and determine if they're allies or enemies of your garden. Their colors, shapes, and sizes are all distinct. Looking at them through a magnifying lens can be fun, but it's not necessary; they are all easy to recognize.

Use the WALA WALA technique two times a week. Damaged leaves and fruit will stand out—they don't require binoculars or a search light. Spend a few minutes looking at your tomato plants as individuals: *glance* at the top and bottom of the leaves for any damage, *scan* fruits for damage, take note of the insects you do see. Again, damaged leaves or fruit will catch your eye. You just have to make a habit of WALA WALA and your enemies will be revealed to you. *If* they are there.

Take a few notes and/or photos of any damage or insects you see. In my experience, the majority of tomato gardeners only deal with two, maybe three, different insect pests that amount to anything. You don't have to learn to identify every possible insect in your garden. It can be fun to know them all, but it's not necessary.

Compare what you see with the photos in this book or other resources, like handouts from your local university agricultural extension office.

Recognize that no one book can fully cover every possible insect, so ask experienced, local gardeners for permission to send them photos of what you're finding if you can't pin down the critter's secret identity on your own.

My bottom line: I've grown tomatoes at home for over forty years. I've rarely had to spray for anything. Ever. Like I said earlier, spraying is not gardening. When I do see insects or signs of their damage, I just smoosh them while I'm there. I also tear off the leaves or fruits they've damaged or been found on—in case they already laid eggs—and toss them to the chickens or the compost.

Last one, and a very important one at that. I've said it before, and I'll say it again: *Too much nitrogen* will make your tomatoes grow faster, but it will also *undermine their immune system and make their leaf color more attractive to scouting pests*. Don't use more than the amount of fertilizer recommended on the bag or on your soil test. A good amount of compost, adequate organic fertilizer, steady irrigation—but not too much—and a layer of organic mulch that's no thicker than a brick will create a healthy soil that grows a healthy plant. That's the best thing you can do to minimize the number of insect pests that find their way to your tomato garden.

A wheelbug—so named for the hump on its back—has a lance-like mouthpart underneath its head. When it sees a soft-bodied insect it swings the lance out stabs its prey and sucks out its juices.

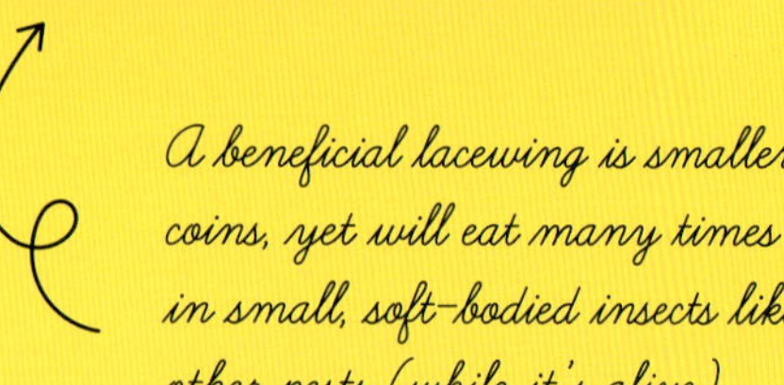

A beneficial lacewing is smaller than most coins, yet will eat many times its weight in small, soft-bodied insects like aphids and other pests (while it's alive).

Juvenile assassin bugs searching for aphids on a grape tomato plant.

An adult assassin bug—of which the wheelbug is another type—also has a lance under its head for sucking juices out of its prey.

We see two different varieties of ladybugs: one with spots and one without. But no matter, these two really "like" each other.

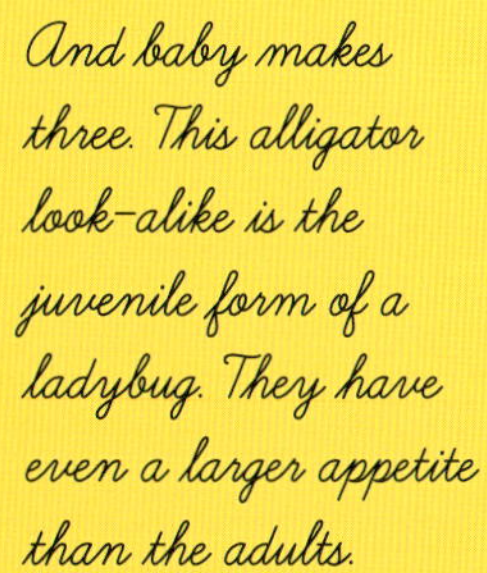

And baby makes three. This alligator look-alike is the juvenile form of a ladybug. They have even a larger appetite than the adults.

Reptiles can also be allies. Tree frogs and toads eat a lot of insects. Non-venomous snakes like this rat snake will eat small rodents like voles, moles, and mice that can nibble on plants.

Name That Pest

As a twenty-something integrated pest management (IPM) scout, I walked miles of rows of lush tomatoes for seven farmers in my home county of Beaufort, South Carolina. My goal was to locate and identify the kinds of bugs and diseases that would harm the plants. The purpose of my work was to help reduce the amount of insecticides and fungicides that farmers sprayed. The pesticide companies encouraged the farmers to spray for every possible problem on a weekly basis: a big cost in time and labor for struggling farmers.

One day, just before tomatoes were ripe enough to pick, I came back from scouting one farmer's 20 acres (8 ha) of tomatoes and reported that I hadn't seen any problems in his fields. To my surprise he said, "There's Colorado potato beetles out there. Gonna spray in the morning."

Wow, was I that bad of a scout? I didn't want to argue with someone twice my age, so I offered to revisit his fields and, perhaps, narrow the range of these hungry insects.

An hour later and I still hadn't seen a single potato beetle (*Leptinotarsa decemlineata*). I was frustrated and doubting myself. The only insect I had seen were numerous lady beetles (*Harmonia* sp.).

Wait. No.

LET'S *do a* DEEPER DIVE

Are you fascinated by bugs and want to know more? Probably the most concise and effective garden insect identification book I've seen is horticulturist Jessica Walliser's *Good Bug, Bad Bug.* (Full disclosure: She's the editor of this book, but I've consulted her book since it came out in 2008, long before we worked together on this book.) Its water-resistant pages mean it can be taken outdoors in poor weather. It shares organic means of control and is chock-full of photos.

Probably the best known beneficial insect is the ladybug. They eat aphids and other soft-body pests.

This experienced farmer couldn't be mistaking these two distinct insects, could he? Potato beetles are twice the size of ladybugs, are a dull yellow color, and have black stripes. Ladybugs are bright red with black spots. But I'd walked his land twice, so I had reassured myself that these fields needed no spraying. I didn't want to tell this older man and experienced farmer to his face that he was wrong. I came up with a plan that might resolve this with little drama and headed back to his garage, where he was wrenching on his tractor.

I told him I still hadn't seen any potato beetles and then held out to him a glass vial a little bigger than a wine cork. I had captured a few ladybugs and put them inside. He pointed at the vial and said, "Potato beetles. Gonna spray in the morning."

I asked if he had the university's handouts with photos of good and bad insects. He took me into his office. When I saw the look of recognition on his face, I silently declared, "mission accomplished" and backed out to get in my truck without adding to any embarrassment he might feel.

Also, almost every county, state, or province in the English-speaking world has a university-based agricultural extension office that can provide handouts to help identify insect allies, pests, and diseases specific to your region. Often these offices are staff by trained volunteers who can help you. Consider them . . . well . . . allies.

The handiest one-page guide for identifying good and bad bugs comes from Mountaineer Books at www.mountaineers.org/books. Search for their Mac's Field Guides to Bugs and choose your region (or your sister region).

CHAPTER 7

More Than One Way to Water

I'VE ALREADY TOUCHED ON THE IMPORTANCE OF KEEPING YOUR TOMATO FOLIAGE DRY WHEN IRRIGATING TO PREVENT THE SPREAD OF DISEASE, AND I'VE EVEN BRIEFLY INTRODUCED A SYSTEM OF IRRIGATION KNOWN AS THE SOAKER HOSE, BUT IN THIS CHAPTER I SHARE TIPS FOR INSTALLING AND USING SOAKER HOSES, AND KNOWING HOW LONG TO RUN THEM. I ALSO INTRODUCE SOME OTHER UNIQUE WAYS TO IRRIGATE YOUR TOMATO GARDEN, INCLUDING SEVERAL THAT COLLECT AND DISTRIBUTE RAINWATER TO KEEP YOUR TOMATOES HYDRATED AND HAPPY.

Soaker Hoses for the Win

Soaker hoses are a great tool for conserving water, whether from the spigot or a rain barrel, but watching novice gardeners attempt to lay out a brand new soaker hose looks like a game of Twister played with a boa constrictor. The stiff, springy serpent wraps around a leg instead of a plant and won't let go. The more they pull, the knottier the problem gets. Sound familiar? Even experienced gardeners can find new soaker hoses frustrating to work with. After decades of gardening with soaker hoses, I've found a few tricks to tame the twisty beast.

1 UNROLL

The manufacturers rolled up the soaker hose into a coil at the plant. So, your best bet is to *un*roll it while walking backward to keep it from getting tangly and twisty. Put a brick or stone on top of the end of the hose to hold it in place as you unroll it down the driveway or across the lawn. It will lay pretty flat and, after few minutes in the sun, it'll be not much more springy than an old garden hose.

2 PIN IT DOWN

Before mulching the bed, put the female end (or call it the "top"—the other end is the "bottom" of the hose) at the edge of the bed, or wherever it will connect to a regular garden hose. Pin the end in place with metal landscape staples. These are *U* shaped and about 3 inches (7.62 cm) long. I find I can hook about twenty of them through a belt loop at my hip on my pants or shorts to make it easy to carry them. If the hose is to be placed in a straight-line row of tomato plants, you only need to pin it down with a staple every 10 feet (3 m) or so. If the soaker hose will be swirling around an ornamental bed of mixed vegetables or landscape plants, you'll need to pin the hose down every time you change direction.

3 LAY AND LOOP IT

The larger the plant, the more water it will need. If you're snaking through a new, freshly planted vegetable bed, loop the hose once around each plant. Lay the hose around the root ball without touching the stem of the plants. Why? If the hose is touching the stems, it may hold moisture there for too long and cause rot. On the other hand, if the hose is too far away from the root ball, the plants won't get enough water. The water from the soaker hose mostly travels down and may not travel very far horizontally—perhaps only a few inches to either side, depending on the slope and the soil type.

4

WATER TEST (THE BEAUTY PART)

After laying out the hose and pinning it down as needed (some spots may still be too springy to be held by a staple, so use a brick temporarily until the hose relaxes), hook it up to a garden hose and turn on the water to make sure every plant's root system is getting watered. Adjust the placement if necessary. Sometimes, it's easier to move a plant than to reposition the hose. This is my favorite part of installing a soaker hose. Why? When the water droplets start coming out of the hose, they catch the sun and look like quicksilver. And you can watch the water as it progresses from the top to the bottom of the hose. The water looks alive and beautiful. When you're happy with the layout, apply mulch. Soaker hoses should be under mulch, but never under the soil. Why? The pores will clog up quickly.

5

HOW MUCH WATER PRESSURE?

Often, the directions on the soaker hose packaging suggest turning the spigot handle only a quarter turn. But I work in older neighborhoods (fifty to one hundred years old) with old, galvanized pipes so that doesn't deliver enough pressure. I start with the spigot opened all the way and that usually works fine, even at a house with new plumbing—but use your judgment. Every situation will vary. Also, some soaker hoses come with a blue "pressure reducer" instead of a gasket in the female end. All my colleagues and I have come to the same conclusion: replacing that with a regular gasket works better, meaning it delivers more water effectively. But YMMV (Your Mileage May Vary).

HOW LONG TO WATER?

I normally use a timer on the spigot; that way a gardener can turn on the water on their way to work and not have to worry about coming home to a flood. They may still worry about having left the oven on—can't help you there—but at least the water will turn off. In our Piedmont clay-based soils in summer, for a newly planted bed, I recommend running the soaker hose for two hours, two times a week if there's no rain. But that's just a guideline. The gardener has to determine how much to water by experience—if plants are wilting in the morning, increase the watering. If areas are boggy, cut it back. Since plants are like cold-blooded animals, when cool weather comes, the plants need less water. When frost threatens, empty the timer and bring it indoors so it doesn't freeze and burst. The soaker hoses won't be bothered by the cold.

7

HOW MANY HOSES CAN YOU STRING TOGETHER?

I've strung as many as three hoses, each 75 feet (22.86 m) long, together. Your results may differ depending on conditions and your water pressure. The key is to have the water entering the soaker with the "top" end of the hose uphill from the "bottom" end. Why? As soon as the water enters the soaker hose, it starts to lose pressure. So, if the soaker hose were running *uphill*, much water and pressure would be lost at the uphill end, leaving those plants too dry and the bottom plants too wet. With the soaker hose running *downhill*, the water will be distributed more evenly, even if the hose is running across the slope for much of that run.

8 LONG-TERM ISSUES

Soaker hoses are cheaper to install than drip lines and so there is a trade-off. After about a year, soakers only deliver about half as much water due to their pores clogging up. They are still effective, but just less so. Sometimes, a critter will bite a hole in the hose to get the water, or after a few years a crack will open up. You can cut out this damaged part with hand pruners and insert a store-bought "hose mender" that splices the two ends together.

9 THE BOTTOM LINE

Soakers will save money and time on watering, reduce weed pressure, recycle waste material (some brands are made from old tires), can be repaired by average gardeners, and last long enough to sustain many organic gardens. And they don't cost much and are very easy to install once you know how to wrestle the beast!

A battery-powered timer lets you set a schedule for watering. Even when you're on vacation, the garden will get the water that it needs. A "Y" valve attached to the spigot lets you install a separate hose for hand watering.

Keep an eye on how much water your garden is receiving with an easy-to-read rain gauge from your local garden center.

This is the least-cost option for storing a garden hose. Especially since this one is a thrift-ique shop find. It's also wall-mounted near the spigot.

Speaking of spigots, double-check to make sure the water isn't dripping, or you could get an upside-down icicle like this.

The least-hassle way to store a garden hose is with a wall-mounted metal spool, placed near the spigot. The hose will be easy to roll up and store, as well as easy to unroll and put into action. Plastic spools will break in a few years. Spools mounted on plastic wheels on the ground are a royal pain to maneuver, spool, and unspool. In a sane world, they would be outlawed.

Water Torture *from* Leaky Hoses

It's rare that I visit a client's or friend's garden and don't see a leaky hose connection. Often, someone has used a wrench in an unsuccessful attempt to squeeze out the drip, but they've merely managed to make the hose nearly impossible to unscrew by hand. Ninety-nine percent of the time, a leaky hose connection is due to a missing or worn-out rubber gasket. Replace the gasket and stop the leak. No more wasted water, no more annoying drips. Any garden center or hardware store can sell you a little container of gaskets for a few bucks. Keep them handy, and stick a few in your pocket when you visit your gardening friends. When you fix their leaky hose, they'll think you're a saint. Or a genius. Or both. They'll probably give you a few bags of zucchini to take home.

Gaskets are inexpensive. Which is probably why they fail every so often. Keep some handy.

LET'S *do a* PROJECT

Water Your Tomatoes with a 300-Gallon (1,135.62 L) Rain Barrel (That's Also a Water Garden!)

When other gardeners installed rain barrels made from 55-gallon (208.19 L), plastic pickle barrels, I wanted in on the fun too, but I started looking for something that would be bigger and look better. At our local feed and seed store, I found what I was looking for: a 300-gallon (1,135.62 L) galvanized horse trough. (Your local garden center may be able to special order and deliver one. They come in different shapes and sizes.) With a downspout diverter from a tool supply catalog, I was able to pipe rainwater into the trough, and since the trough comes with a threaded opening near the bottom I could add a brass spigot and a soaker hose to water the nearby vegetable garden.

I wondered if I could do more with this trough. Keep fish? Grow edible aquatic plants? That was about ten years ago and I now have two of these at my house and have installed several for clients. Here's what to do if you want one.

My rain-collecting horse trough doubles as a home for blooming and floating aquatic plants.

The galvanized coating will protect the steel from rusting, but not if it touches the soil where organic acids will oxidize it. If you can't install it on a patio, then raise the trough off the ground on pressure-treated 6 × 6s (140 × 140 mm) or bricks or cinderblocks and make sure it's level.

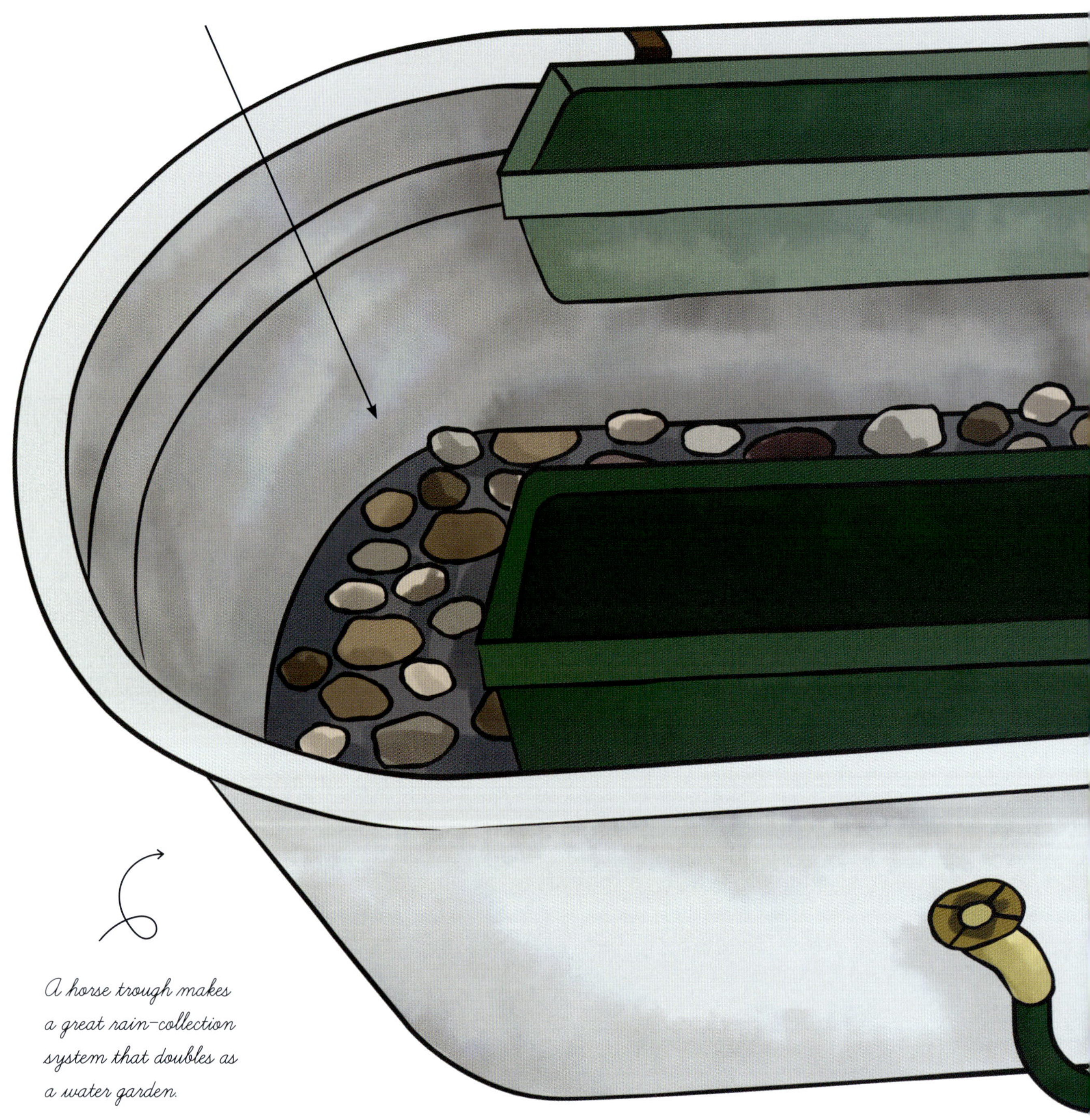

A horse trough makes a great rain-collection system that doubles as a water garden.

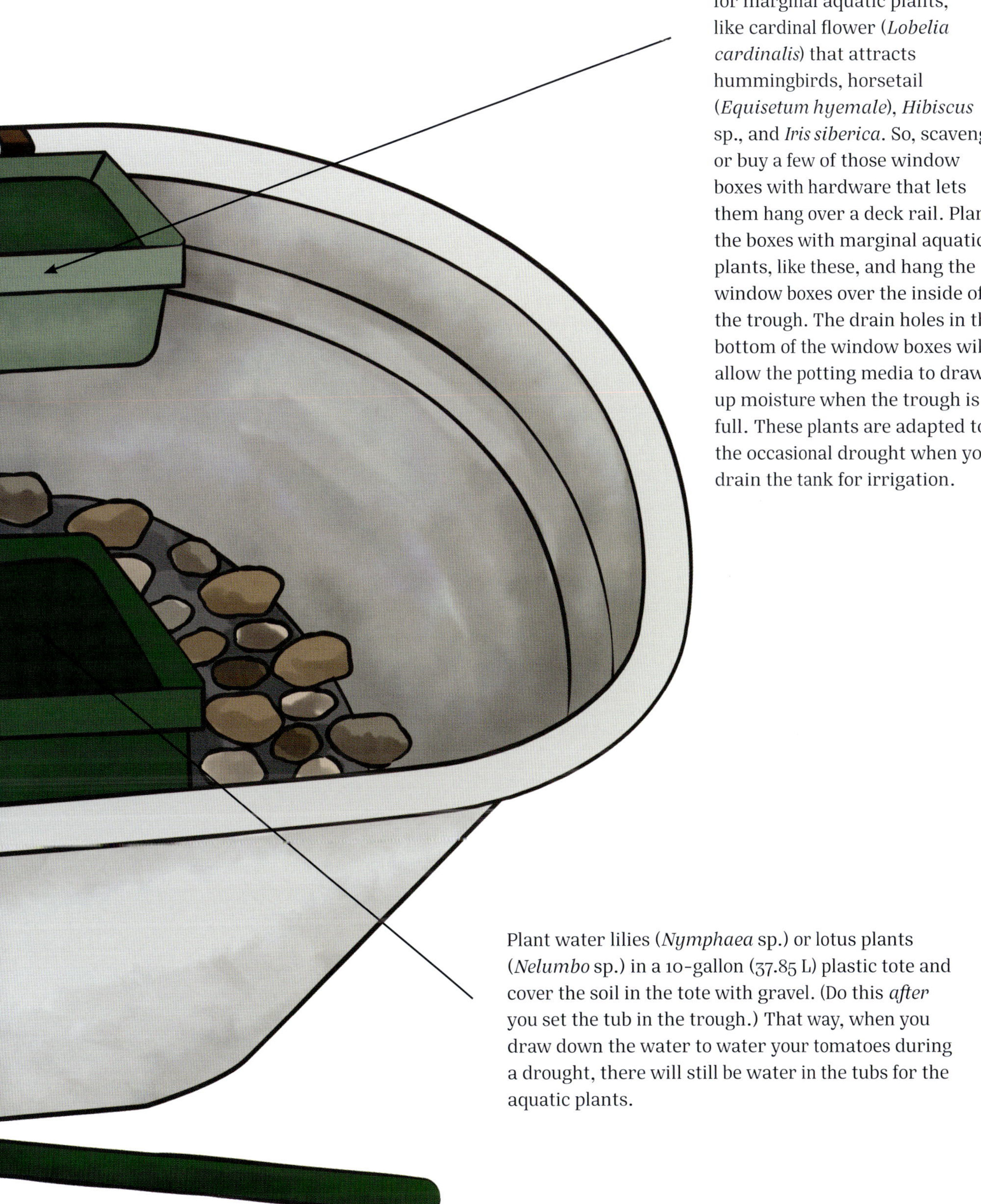

The trough will be too deep for marginal aquatic plants, like cardinal flower (*Lobelia cardinalis*) that attracts hummingbirds, horsetail (*Equisetum hyemale*), *Hibiscus* sp., and *Iris siberica*. So, scavenge or buy a few of those window boxes with hardware that lets them hang over a deck rail. Plant the boxes with marginal aquatic plants, like these, and hang the window boxes over the inside of the trough. The drain holes in the bottom of the window boxes will allow the potting media to draw up moisture when the trough is full. These plants are adapted to the occasional drought when you drain the tank for irrigation.

Plant water lilies (*Nymphaea* sp.) or lotus plants (*Nelumbo* sp.) in a 10-gallon (37.85 L) plastic tote and cover the soil in the tote with gravel. (Do this *after* you set the tub in the trough.) That way, when you draw down the water to water your tomatoes during a drought, there will still be water in the tubs for the aquatic plants.

Make sure your trough is level before adding anything to it.

A horse trough can help water a garden and be a water garden. A tub holds a water lily and window boxes hold marginal aquatic plants.

A 300-gallon horse trough also works as a water garden with goldfish, flowering water lilies, and floating plants.

The troughs come with a threaded opening, so you can screw a brass spigot into place. Attach a hose here that leads to a soaker hose in your tomato garden.

After ten years or so, the galvanized trough may spring a leak no matter what you do. One way to extend its life indefinitely is to cover the inside with rubber pond liner. Clip the liner in place at the rim by cutting 1-inch (2.5 cm) diameter flexible plastic pipe into 3-inch (7.6 cm) long sections and slice them in half. Spread them open and clip them onto the rim.

The window boxes can hold marginal aquatic plants like cardinal flower. The bright red flowers attract streams of hummingbirds through the summer.

The plants in the water garden make a place for spiders to build webs and catch problematic insects.

Of course, you don't want your new water garden/rain barrel to populate your yard with mosquitoes, so during the warm season, add mosquito dunks once a month—they contain a bacteria that harms only mosquitoes. Or, if you wait until algae has started growing on the inside of the trough, you can stock the trough with goldfish from a pet store. Don't buy the expensive ones as they aren't tough and may get eaten by herons or other predators anyway. In spring, buy what are called "feeder fish." These are the tough, inexpensive goldfish that can survive less-than-ideal conditions. Start with five: They will likely multiply. And, here in our North Carolina garden, they go dormant in winter at the bottom of the trough and re-appear every April when it warms up.

Two unexpected bonuses. First, in spring we hear songs of four different species of tree frogs coming to lay their eggs. Second, the afternoon sunlight bounces off the surface of the water up through our living room window and shimmers on the ceiling as you've probably seen sunlight do under a bridge. Can your rain barrel do that?

My rain-collecting horse trough is much more attractive than a plain rain barrel.

LET'S *do a* PROJECT

Water Your Tomatoes with a Box of Rain (a.k.a. a Rain Cube)

BY SUMMER, A VEGETABLE BED NEEDS ABOUT 1 GALLON (3.78 L) OF WATER PER SQUARE FOOT (0.09 SQ. M) EVERY WEEK. WATERING A SINGLE 4 × 12-FOOT (1.21 × 3.65 M) VEGETABLE BED WOULD DRAIN A TYPICALLY SIZED RAIN BARREL OR UPCYCLED PICKLE BARREL VERY QUICKLY AND THEN YOU'RE OUT OF WATER UNTIL IT RAINS AGAIN—UNLESS YOU SET UP A WHOLE SERIES OF RAIN BARRELS OR PICKLE BARRELS. BUT LINING UP MORE OF THEM DOESN'T MAKE THEM ANY LESS UGLY, AND IT MULTIPLIES YOUR COST OF CONNECTIONS AND THE TIME TO INSTALL THEM.

Asking around to see if anyone had a better option led me to IBC: international beverage containers. You can find them advertised on Craigslist for $75 to $150 apiece, a price they are well worth. I figure if it holds more gallons than it costs in dollars, you're getting a good deal. And IBCs hold 275 gallons (1,040.99 L)—more than five pickle barrels' worth—but with a smaller footprint and requiring less hardware.

IBCs are cubes (48 × 48 × 40 inches, or 1.21 × 1.21 × 1 m) of food-grade plastic mounted on a pallet. They make me think of the Grateful Dead song "Box of Rain." An aluminum cage keeps the plastic cube from collapsing. At the top is an 8-inch (20.32 cm)-wide screw-on lid. Near the bottom is a 2-inch (5 cm) spigot. The ones I've bought still had the faint scent of the almond oil they originally contained.

To keep algae from growing inside, I slap on a couple coats of dark brown paint made for plastic. Then, it looks like a cubic Hershey's Kiss. Plus, I install them in a way that they can send water uphill to the garden without a pump. Here's how.

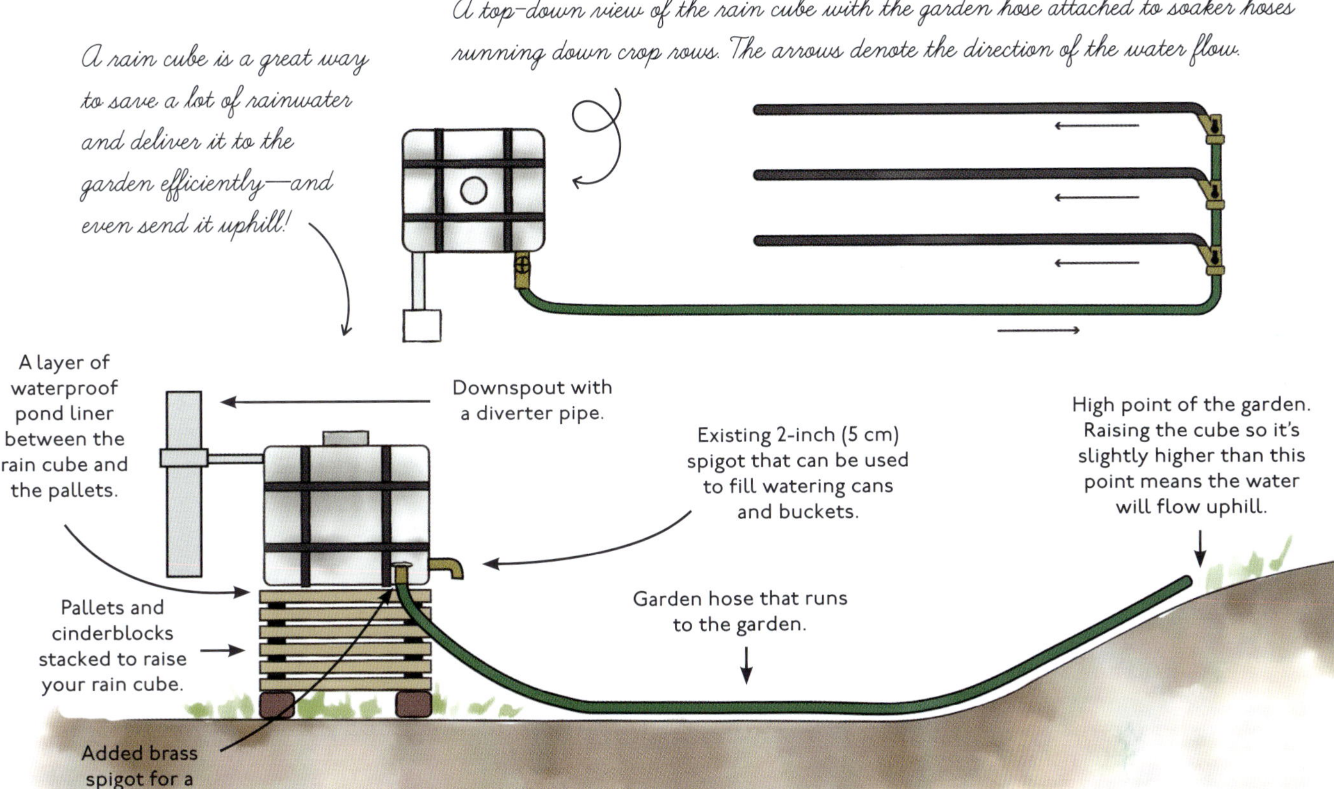

SUPPLIES NEEDED

- Twine
- Line level
- 4-foot (1.21 m) level
- Blocks of 6 × 6 (140 × 140 mm) pressure-treated wood, cinderblocks, or bricks
- Pallets
- Rubber pond liner, metal roofing, or rolled roofing
- Brass spigot
- Spade bit that matches the inside diameter of the spigot's threads
- Exterior-grade caulk
- Downspout diverter and watertight connection
- Spade bit that matches the inside diameter of a watertight connection
- Overflow hose
- Watertight connection
- Exterior-grade paint for plastic

CONNECT TO THE DOWNSPOUT

There are two ways to get rainwater into your box of rain:

1. **Under an existing downspout.** Cut a hole in the top of the box and drop in the downspout. Use a section of downspout to mark an outline on the top of the box. Drill a hole just inside the line. Insert the blade of your jigsaw into the hole. Cut along the line and your downspout should fit snugly through the hole. You'll need to add the same kind of overflow hose near the top of the box that any pickle barrel or store-bought rain barrel would have to divert excess rainwater when the box fills up.

2. **Not under a downspout.** I recommend using the downspout diverter sold in the Gardener's Supply Company catalog. It's a metal box that fits into the downspout. It diverts water into a hose that connects to your rain barrel. Install it correctly (meaning that, even if you think you know how, you read and follow the directions that come with it) and you won't need to mess with an overflow hose. When the box of rain is full, the rainwater will back up into the hose and go down the downspout instead. Installing this device correctly means the inlet to the box is at the same height or lower than the diverter. If you want to set up your box farther from the downspout, then the Gardeners Supply hose allows, buy ¾-inch (1.9 cm) plastic hose used in Netafim drip irrigation systems. These are often sold at big-box stores. Just be sure not to use a clear plastic hose. Algae will grow in it, making it ugly and clogging it up.

BUILD A BASE FOR THE BOX OF RAIN

Your box of rain comes with one pallet. I think it works best stacked on several more pallets so you can set a bucket under the spigot and send water uphill (more on that later). Get free pallets from feed and seed stores, garden centers, HVAC contractors, big-box stores, etc. If they have "HT" stamped on the side, they have been heat treated and don't have chemicals in them. Most pallets that have been treated with chemicals have bright paint slapped on the side and are expensive, so stores send those back for credit.

Once you have all the pallets you need, raise them off the ground by putting a cinderblock under each corner. Bricks or scraps of pressure-treated lumber work fine too. The main thing is to get the pallet wood on a rot-proof foundation at least 5 inches (12.7 cm) above the ground so it won't decompose. Level the foundation pieces, then start stacking pallets. Weight and friction will keep the pallets from moving around so you won't need to screw them down. Check that the pallets are level, or pretty close to it.

Then, wrap the pallets and foundation with some fence wire to keep critters from making a home under your box of rain. Extend the fencing as an apron on the ground around the foundation to keep critters from digging their way under the pallets.

Last, use something waterproof and UV-proof as a flat roof to keep the pallets dry: scraps of pond liner, rolled roofing, scraps of metal roofing, etc. Most tarps won't stand up to the sun's UV radiation, so don't go that route. I had just enough scraps of pressure-treated lattice, so I used them to cover the sides of the pallet stack, both to keep critters out and to make it look a little more like a garden feature, but that's not necessary.

Once your pallet stack has a roof, foundation, and critter proofing, put the empty box on top. It should be light enough for one or two people to lift it into place. No need to screw it down. Short of a tornado, its weight will keep it in place.

The base can be any convenient height. Here, four pallets are kept off the microbe-infested soil by bricks that act like a foundation, wrapped in rubber pond liner that acts like a roof and siding to keep rain out and covered with lattice for a finished look.

Add a spigot and hose to your box of rain. The hose can be connected to a soaker hose in the tomato garden. The metal flamingo is what Chris calls "a decorate."

LET WATER FLOW

Some people unscrew the 2-inch (5 cm) spigot and use lots of PVC adaptors to dial down the diameter so a brass spigot can replace it. I tried that once and found that the adaptors seriously constricted the rate of flow. Not only that, but I missed having a big spigot that could fill a 5-gallon (18.92 L) bucket or a 2-gallon (7.57 L) watering can in a matter of seconds. So, I now keep the 2-inch (5 cm) spigot in place and make a level area under it with bricks where I can set a bucket or watering can and fill it quickly.

To water my garden with a soaker hose, I attach a brass spigot near the bottom of the cube. To do this, I use a drill and a spade bit that cuts the *same diameter as the inside diameter of the threads* of the spigot. That way, the outer edge of the threads will bite into the plastic as I screw it into place. Just before it's screwed all the way in, I apply a bead of exterior-grade caulk around the threads and then screw it in the last bit. Leave the caulk to cure for twenty-four hours before moving anything or letting water into the cube.

SEND WATER UPHILL

Here's the fun part. I had a good site for my box of rain, but the garden was uphill from there. Carrying water by hand uphill was out. Installing a pump and routing electricity was expensive, and pumps like to break down anyway.

This was one of those moments when I feel like I learned everything I need to know in middle school science class. I remembered learning that water will run uphill through a hose if the inlet is raised higher than the outlet. This is the same principle that gets water from a city water tower to your house. That's when I realized that if I stacked the pallets just a tiny bit higher than the high point of the garden, gravity would push the water uphill into the garden.

Follow these steps before building your pallet stack and you can do the same:

1. Drive a stake into the high point of the garden and wrap twine around its base at ground level.
2. Pull the twine to where the box will be stationed.
3. Hang a pen-size line level on the twine.
4. With the twine pulled tight, adjust it until it reads level.
5. Measure the distance the twine is above ground level; in my case this was about 35 inches (88.9 cm).
6. If the spigot of your box of rain is a couple inches above that height, the water in the hose will "think" it's going downhill, even if the hose lying on the ground is actually running uphill.

Pallets are about 5 inches (12.7 cm) tall and I had some 6 × 6 (140 × 140 mm) scraps for a foundation. In this case, six pallets put the box's spigot just above the hose outlet at the high point of the garden. The rainwater would flow uphill. From the box to the high point, I would use a garden hose. From there, a soaker hose carried water downhill (remember, a soaker hose running uphill loses most of its water right away).

MAINTENANCE

Your box of rain should be mostly maintenance-free. Close the spigot when it's empty so it will refill during rains. Come winter, open the spigot so water doesn't freeze inside. Depending on conditions, you may have to touch up the paint periodically to keep algae out of the water. Keep an eye out for critters setting up house in the pallet stack if your fencing isn't tight enough. Keep your rain gutters free of leaves in fall and oak flowers in spring (hopefully you're doing that already). Then, enjoy tons of free water from your box of rain.

Save Time and Water with a Timer

Occasionally, I see gardeners who are struggling to keep their gardens watered by hand. Most hoses might be putting out about 10 gallons (37.85 L) per minute. For an urban farmer in an older part of town, corrosion in the old pipes may cut that rate in half.

To grow well in summer, a tomato garden needs roughly 1 or 2 gallons (3.78 or 7.57 L) of water per square foot (0.09 sq. m) per week, if there's no rain. So, to water about 600 square feet (20 × 30 feet, or 55.74 sq. m [6 × 9.14 m]) of garden would take one to two hours of standing around in the hot weather. Every single week. And there's still the rest of the garden to tend to.

But there's an alternative I rely on: Inexpensive timers mounted on the spigot allow the right amount of water to run through a soaker hose and then shut the water off for you.

I know timers work because I set up all my garden clients with them. Since 99 percent of my clients *aren't* gardeners, the timers let me set up a simple push-button irrigation system that allows even nongardeners to keep new plants alive. In thirty years of installing gardens for nongardeners, I haven't had any call-backs to replace plants.

These timers work the same way a kitchen egg timer does. Turn the dial to a set amount of time and—tick, tick, tick—it works its way through a countdown then turns off the water on schedule. They cost $10 to $15 and are available at most garden centers. It's this easy:

1. Screw the timer onto the spigot.
2. Screw the garden hose onto the bottom of the timer.
3. Connect the garden hose to a soaker hose in the thirsty tomato garden.
4. Remove the timers in winter so ice forming inside won't crack them open.

You can do as I suggest to my clients: Turn on the timer on your way out of the house. You can be confident the watering will stop while you're away.

Then, when you come home, if necessary, you can move the garden hose to water another bed and set the timer again before you go in to make dinner. No standing around required.

LET'S *do a* DEEPER DIVE

Add to your irrigation skills with Robert Kourik's reliable book *Drip Irrigation: For Every Landscape and All Climates*. His book provides better results with fewer contraptions. He covers rain barrels and gray water irrigation.

CHAPTER 8

Don't Let the Weather Give Your Tomatoes the Cold Shoulder

SPRING WEATHER CAN FLUCTUATE MORE SO THAN OTHER SEASONS. YOU NOW KNOW TO WAIT TO PLANT YOUR TOMATO TRANSPLANTS UNTIL TWO WEEKS AFTER THE LAST *AVERAGE* SPRING FROST. BUT YOU CAN STILL GET SOME COOL OR COLD-ISH WEATHER EVEN AFTER THAT DATE PASSES. UNDER IDEAL CONDITIONS, WE WOULD ALL LIVE IN REGIONS WITH LONG, BALMY SUMMERS FOR OPTIMAL TOMATO GROWING. BUT IF YOUR SEASON IS SHORT AND WINTER LIKES TO LINGER, THERE ARE A COUPLE OF THINGS YOU CAN DO TO HELP YOUR TOMATOES GET ROLLING IN SPRING. HERE ARE SOME OPTIONS.

Black Plastic Mulch Warms Cold Soil

If you wear black clothes in summer, you've noticed they can warm you under the sun. Dark colors absorb warmth from the sun and lighter colors reflect it. That's why either a black plastic mulch or a black biodegradable mulch helps warm the soil your tomatoes are growing in.

In cool spring soil, an organic mulch, like shredded leaves or bark, keeps the sun from warming the ground and tomato transplants can become listless. But with a shot of solar warmth from a thin layer of black plastic mulch, tomatoes are stimulated to grow a bit faster in spring, even when the weather is slow to warm up.

Black plastic mulch comes in a heavy roll 3 feet (91.44 cm) wide and 100 feet (30.48 m) long. It will fit a bed of soil 18 to 24 inches (45.72 to 60.96 cm) wide, that's raised about 4 to 6 inches (10.16 to 15.24 cm) high. After shaping the bed, put a soaker hose or drip line down it, a few inches away from the center line, where the plants will go. You need to allow a few inches on each edge of the mulch to cover it with soil in a continuous row and hold it in place. If you just dump a shovelful of soil every so often, the breeze will easily flow under the uncovered edges of your black plastic and carry away all the warm air you are trying to generate. (Tossing some soil on top of the plastic every so often to keep it from blowing away as you install it is wise, but come back and add more soil to completely cover the edges.) When each side and both ends are covered in soil, the heat from the sun striking the black plastic will stay in one place long enough to heat up the underlying soil and roots. This will speed up growth, even on cool spring days.

When you're ready to set out your tomato transplants, walk down the row and use a sharp knife to cut out a square that's about 3 inches (7.62 cm) wide: enough room for your hands to plant a tomato transplant. Make sure that neither the soaker hose nor the black plastic touches the stem. The first could promote diseases, the second could "cook" the stem on a hot day.

In addition to warming the spring soil, the black plastic will also hold in soil moisture, suppress annual weeds, and minimize the likeliness of soil-borne microbes contacting the tomato plant's lower leaves.

Nonwoven Polyester Fabric Holds Warmth and Allows Sunlight and Rainfall to Pass Through

Another way to boost spring temperatures and speed up plant metabolism even more is to use white, nonwoven, spun-bound polyester fabric. Sometimes called row cover, Reemay and Hollytex are common brands. This gauzy-looking fabric lets sunlight and rain through to the plants and soil. Plus, it blocks flying insects.

It's not solid, but it does reduce airflow. That's how it acts like a greenhouse to hold in warmth. Placing a layer of this lightweight fabric over your plants *can* create a mini greenhouse. I say "can" because it's also necessary to anchor the entire edge of the fabric so the breeze doesn't get under it and chase off all the warmth you're trying to capture.

This fabric doesn't create warmth—it captures and holds it. Some warmth comes up and out of the soil overnight, and the fabric can capture that. Some warmth delivered by the sunlight would bounce off the plants and organic mulch and drift away on the air, but sunlight has an easier time flowing down through the fabric than it does escaping the fabric. But that's all conditional, of course.

If your fabric is draped over the plants *but not secured all along the edges*, any wind that gets up under it will chill and replace the warm air you've been trying to capture. Kind of like leaving the doors and vents of a greenhouse open on a cold day. That defeats the purpose.

But, if your fabric is pinned down on the edges, then the wind will flow over the fabric, leaving that valuable warmth in place.

When I first confronted this issue, I used scraps of lumber to line both edges and each end of the fabric—but that's a lot of effort and material. Especially if you want to remove the fabric to deal with an issue in the plant bed. Some people use landscape staples to hold the fabric in place, but any spot that's slack will open up and let that cold air blow through.

I decided there had to be a better way to use this fabric that kept the warmth in, but was also easy to remove and replace. Out came the bolt cutters and some scraps of concrete wire—called remesh—I use for making tomato towers. I call the resulting tools "baby greenhouses." Here's how to make them.

BABY GREENHOUSES TO HOLD HEAT

I cut 6 whole squares of remesh with one complete set of "fingers" on each side (more details on using remesh in chapter 5). The remesh roll is 5 feet (1.52 m) high—or long in this application—and I cut it to 36 inches (91.44 cm) wide (6 whole squares) with 6-inch (15.24 cm) "fingers" on each side so the remesh section is 48 inches (121.92 cm) wide after it is cut. Its natural curve from being rolled up serves our needs perfectly. But do make sure it's tall enough—12 to 20 inches (30.48 to 50.8 cm)—for your tomato seedlings to grow up for a couple weeks, until the spring weather evens out.

The view from my roof. Half the beds are covered with baby greenhouses and half left uncovered.

With the remesh cut to size, I lay some row cover fabric—roughly a 5 × 10-foot (1.52 × 3 m) dimension—on the ground and place the remesh on it, "fingers" up. I pull up one side of the fabric to the finger side of the remesh and pierce the fabric with the fingers. Leaving about a 6-inch (15.24 cm) excess.

I then pull up the other side of the fabric and leave enough slack so I can pierce the fabric with the fingers without stretching the fabric. Flip it over, fingers-side down, and you'll see where we're going with this.

Placed over a bed or row that's about 2 feet (60.96 cm) wide, the fingers can be pushed into the ground where they pin the fabric down to the ground in a way that won't allow the wind to get up and under it.

At each end of this baby greenhouse, there should be enough loose fabric to fold together and cover with a scrap of 2 × 4 (38 × 90 mm) or loose bricks that also press the fabric down against the soil to keep the wind out.

These baby greenhouses, situated so the wind can't get under the fabric, will make a difference during the fluctuating conditions of spring, but they work best as a combination that starts with:

- A prepped bed with proper pH, nutrients, organic matter, and loose soil
- A soaker hose or drip line to keep irrigation off the leaves
- A layer of black plastic mulch to keep warmth and moisture in the soil

Push the fabric over the "fingers" of the remesh to hold everything in place. In the off-season, they can be stored outdoors but out of the sun which would damage the fabric.

Once the weather stays warm—and the tomato plants are tall enough to clamor for a trellis—you can easily remove the baby greenhouses and store them away, nested, out of the sun and waiting for fall, when they'll be used again to cover greens and crucifers through the cool part of the year.

A close-up view of the remesh "fingers" carefully pushed through the Reemay fabric.

A baby greenhouse in a spring snow, holding the ground's warmth around transplants.

Dress for Tomato-Growing Success

My life-support system in hot weather. A big fan, a big straw hat, cloth-backed gloves, sweat-wicking clothes and cold drinks.

I remember a time when I can truthfully say I enjoyed sweating. It allowed me to feel the sensation of every square inch of my skin at one time. That amplified my feeling of being alive while working or playing outdoors. Strangely, as I've acquired an appreciation for comfort, working during the sweaty season has lost a bit of its appeal.

I still enjoy the sensation of feeling alive and the feeling of being a part of nature I get from outdoor work. But being a practical guy, I've also adopted a few techniques and tools to get me through sweat season with comfort.

Working from the outside in, I've spent about $100 for a darn big fan. It has a 24-inch (60.96 cm) diameter, a handle, a long cord, an adjustable stand and two settings: low speed and loud, or hurricane speed and loud. I have never minded the loud part. It's especially helpful when I'm working in one spot—building a raised bed, potting up plants, etc. It makes a sunny, hot, summer day feel like a sunny, cool, spring day. I have it on even when I'm moving around doing some planting, pruning, or weeding. When I need a break, I plop myself in front of this air blaster. It's almost as refreshing as dumping ice water on your head, but without the part where your heart comes to a full stop. (But if the heat does have you feeling mentally foggy, then a big splash of cold water on your head and down the back of your neck is the exact right thing to do.)

I've also adopted the head gear of the universal peasant: a wide-brimmed straw hat. If I'm driving around, I loop the neck strap over the headrest of the passenger seat with the hat hanging behind the seat where it's out of the way but easily accessible. A wide brim not only protects your face and neck, but also keeps the sun off of most of your shoulders. It's worth searching out a hat with as wide a brim as you can find. And keep your hat dry, but don't let it get too dry. That's when the fibers start cracking and breaking apart. Maybe once or twice a year, I leave it outside in the rain to revive by soaking up some moisture. Or use it as a rain hat in a light rain. Or push it underwater in the rain barrel for just a few seconds. Whichever way you wet it, let it dry out slowly in some shade. Sunlight will dry it too quickly and will defeat your purpose. The fibers need a very small percentage of moisture to stay limber.

With or without my straw hat, I often wear a headband that has what's called a sweat gutter. It doesn't look like a rain gutter on your house. It's just a thin strip of flexible plastic that runs above my eyebrows. The sweat that would normally run down my forehead and get into my eyes or on my glasses—even with a regular sweat band—reaches that thin gutter and flows to the right or left, away from the front of my face. The first time I tried it, I was dubious—

but they work. I bought several more so there would always be one in each vehicle and on the coat rack by the backdoor. And a few to give to colleagues.

Generally, I'm not a fan of plastic clothes, but I've been making good use of CoolMax brand T-shirts and boxers. They don't get soggy and clingy like cotton clothes and they help your sweat evaporate like it should, to cool you down. Their one downside is they can retain some body odor. But I've found a solution to that. When I run them through a load of wash, I add ¼ cup (60 ml) of antibacterial hand soap to kill the stinky bacteria. It works, and since that's my only use for that kind of soap, a refill bottle lasts a long time.

Of course, our body gets rid of heat through our extremities, so wearing gloves can leave you feeling hot. The concrete supply house and an old-fashioned hardware store in our community sell cotton gloves with a waterproof layer on the palms in packs of ten pairs for $30 or $3 a pair: cheap enough that I can leave a couple pairs in all the different work stations on our homestead. I can handle wet or rough material with the palms and fingers of my hands protected, but the back of my hands and fingers can breathe through the cotton to keep me from overheating.

On rainy days, I wear muck boots, but the black ones get very hot in summer when the sun comes out. The solution is to buy white muck boots that shrimpers and other professional fishers wear. Joy Fish is a good brand. They are just as waterproof as the black muck boots, but the white color reflects heat instead of absorbing it.

Staying hydrated is important too, so I keep a non-BPH water bottle handy with a carabiner that runs through the loop for the cap. I can clip the carabiner onto a belt loop to keep it handy. Have you ever noticed that you can drink plenty of water on a hot day and still feel very fatigued by day's end? That's because you've sweated out a lot of salts that haven't been replaced. I've never liked the flavor of sports drinks—I think they add lots of unnecessary sugar to simply obscure the salt. Yuck. Instead of that, I suggest countering the loss of salt by starting the day with a can of V8. I like the flavor of all the vegetables, and the big shot of salt—which would be too strong under other circumstances—carries me through the day. There's a real difference between the way I feel after a sweaty day with or without a V8. I buy it by the case as soon as the weather warms up. One sunny week, I ran out of V8. I remembered my childhood judo teacher made us all take a salt tablet before class. I had no salt tablets. So, to help me get through the day, I wetted a finger, put it in the salt cellar on our table, and put the salt on the center of my tongue. There are fewer taste buds there so the flavor didn't overwhelm me. (And I use a table salt in which half of the sodium has been replaced by potassium. It's healthier and tastes the same.) I got through the sweaty day feeling fine. I stopped buying V8 after that.

And then at the end of a really hot day, I break out a beer. But only to apply the ice cold can to my forehead and the back of my neck, of course.

Your Garden Has a Sister Region (That Probably Isn't the One Tomatoes Come From)

Why do gardeners in some parts of the world struggle to get tomato plants to survive the growing season while tomatoes in other places thrive with little trouble?

It has to do with a concept called "sister regions." I learned about this concept from my mentor Dr. J. C. Raulston. In the way that some sisters can wear the same *pants*, sister regions can share the same *plants*. The corollary being that when shifting a plant—such as the tomato—from the region in which it evolved to a region with different climate and soils, you may struggle to keep that plant happy. But once you understand the sister region concept, you will have an easier time adjusting your conditions so you can grow tomatoes—and many other tricky plants—just about anywhere.

I conclude this chapter with a description of the tomato plant's native region, since it's a low-population region compared to most others. Once you're familiar with your sister region, you'll better understand what your tomato plants need from you. Know you're not alone if you've been struggling to keep tomatoes alive and healthy all season.

The sad end of the tomato-growing season. Be ready to gather the last of the tomatoes before the first frost shows up if you live in a cold region.

How about an example of sister regions?

EAST MEETS EAST

Located in North Carolina, my garden and others on the East Coast of North America share some general qualities: annual rainfall is about 40 inches (101.6 cm), which means there's lush plant growth, which means there's a good amount of organic matter in the soil, which means the native soil has an acidic pH of about 5. With me so far? If not, this short paragraph is well worth a reread.

Interestingly, the same conditions of climate and native soils largely apply on the east coast of China, Korea, and Japan: annual rainfall of about 40 inches (101.6 cm), lush plant growth, high soil organic matter and acidic pH of about 5.

Conditions in these two parts of the world are so much alike that the native plants of one region thrive—and can even become invasive—in its sister region. East Asian kudzu, wisteria, privet, and honeysuckle are exuberant and invasive in eastern North America. Goldenrod and hydrangeas native to eastern North America spread madly along eastern Asia's roadsides.

But it's not all bad news.

If you've been to the Cherry Blossom Festival in Washington, DC (38 degrees N latitude), you've witnessed a striking example of sister regions borrowing each other's plants. In 1912, the Japanese government in Tokyo (35 degrees N latitude) gave three thousand cherry trees to the United States as a gift of friendship. Conditions in the two areas are similar enough that the nonnative cherry trees perform well in the United States. In fact, the average peak bloom dates are nearly identical: April 4 in DC and a day later in Tokyo. So, knowing one's sister region can be a boon for gardeners when it comes to selecting plants that will grow and thrive.

But if you follow that line of latitude east, from DC to Europe, you find yourself in the Mediterranean, where it hardly rains all summer. Plus, the pH is an alkaline 8. That would deprive Japanese cherries of the soil nutrients they need. So, if this concept isn't just about latitude, what is creating those sisterly conditions between eastern Asia and eastern North America?

Unlikely as it might sound, the ocean currents determine where sister regions exist.

Eastern North America's weather—and soils—are curated by the Gulf Stream which draws warm water, warm air, and humidity from the tropics north to sideswipe North America.

On the other side of the world, the Kuroshio Current in the Pacific Ocean performs a similar role along the east coast of Asia, sowing warmth, acidic pH, rainfall, and humidity on coastal China, Korea, and Japan.

THE CAFFEINATED COASTS

The Atlantic's Gulf Stream continues northeast, away from North America and curves clockwise down the coast of western Europe. Regions there, at the same latitude as Canada—the British Isles, Norway, northern France, and northwestern Spain—have much milder winters, even at that high latitude, because the Gulf Stream is still warm.

The Asian Kuroshio Current takes a similar route, northeast away from Asia and clockwise across the Pacific as it meets the western shore of North America. As it curves to the south, it moderates the climate of the region from Juneau to San Francisco.

Both areas—coastal northwestern Europe and coastal northwestern North America—have mild winters for their latitude. Because the warmth of the respective currents have picked up moisture as they cross their oceans, they brush each coast with misty, drizzly rain and lots of overcast days blocking the sun. Is it any surprise that England's love of caffeinated tea is matched by Seattle and Portland's addiction to coffee?

And where in North America can you find easy-to-grow perennial borders reminiscent of an English garden? In the region, on similar latitudes, we call the Pacific Northwest.

With that steady precipitation, both regions have rainfall of 30 to 60 inches (76.2 to 152.4 cm), lush growth, high soil organic matter, and an acidic pH of around 5.

In all the regions described in this chapter, there will be local exceptions to these patterns. For example, the major cities of the caffeinated coasts—London, Dublin, Seattle, and Portland—get much less rain because they are in the "rain shadows" of mountains and high ground to their west. When the prevailing winds from the west push moisture-laden clouds up and over a mountain range or other high ground, the clouds dump their water on the western side of the mountains as they rise, creating wetter conditions there. So, the west side of the Cascade Mountains of the Pacific Northwest and the west coasts of Ireland and Britain are notoriously wetter. And as the now almost-rainless clouds descend over the trailing side of the mountain range, there will be drier conditions to the east. This drier side of a mountain range is what's called the "rain shadow." And, often, larger cities grow there to avoid those damp conditions.

So, yes, there will be exceptions to some of what I describe. But you know the old saying: "It's the exceptions that *prove* the rule."

WINE COUNTRY IS ALSO TOMATO COUNTRY

Continue south of the caffeinated coasts on these ocean currents and, on both continents, you'll find the "Mediterranean"-like climate of southern California and, at a similar latitude on the other side of the world, its sister region, the actual Mediterranean.

The Mediterranean regions all have about 20 inches (50.8 cm) of rain per year. Almost all of it in fall, winter, and spring. Summers are notoriously dry. That means sparser growth—yet more than in a desert with less than 10 inches (25.4 cm) of annual rain. So, lean soils with little organic matter and a higher pH of about 8 are the norm.

FLYOVER COUNTRY

The Great Plains of North America, the Pampas of South America, the Steppe of Eurasia, and the savannas of Africa and Australia are all far enough from the moderating influence of the oceans that horticulturists call their atmospheric conditions "continental weather," meaning summers and winters are often long and harsh. That's a combination that keeps the populations of two-legged creatures low and the populations of four-legged creatures high.

Rainfall varies greatly, but is often between 40 and 20 inches (101.6 and 50.8 cm), creating a neutral-ish pH of around 7. There's rarely enough rainfall for forests, but plenty for grasslands for grazing and grains for harvesting.

A GIANT MAP OVERLAP

If you overlap a map of the east and west coasts of the Eurasian continent over the east and west coasts of North America and align their latitudes and ocean currents, you almost have a paint-by-numbers map of where Eurasian exotics would likely perform well in North America and vice versa (or where they'd need extra attention, such as additional lime for Mediterranean plants grown in the "nonsister" eastern North America).

And, at the center of each continent, you'll find similar harsh weather, unmoderated by ocean currents. Not surprisingly, these areas are subject to that "continental" weather: hard summers, hard winters, and severe periods of drought. In Eurasia, they call it the Steppe. In North America, the Plains. Sister regions.

Obviously, features such as mountains can tweak these patterns; mountain ranges in the US Northwest can cause rain shadows that create nearly Mediterranean-like conditions near Portland, Oregon, for instance. But having this basic understanding about climates, regions, and the evolution of plants, can help gardeners and horticulturists anticipate a plant's likelihood of success in their backyard.

Map Key

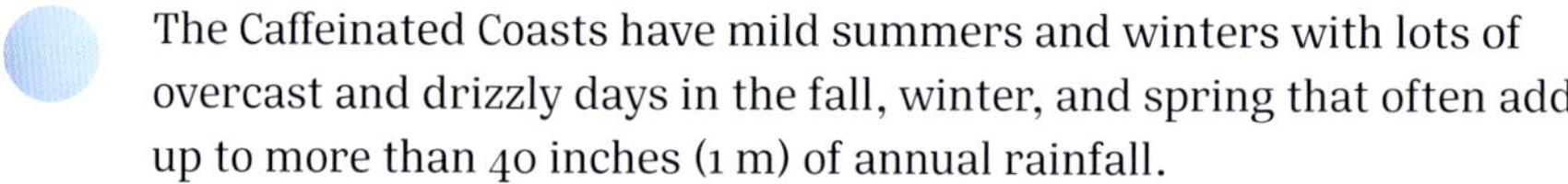

The Caffeinated Coasts have mild summers and winters with lots of overcast and drizzly days in the fall, winter, and spring that often add up to more than 40 inches (1 m) of annual rainfall.

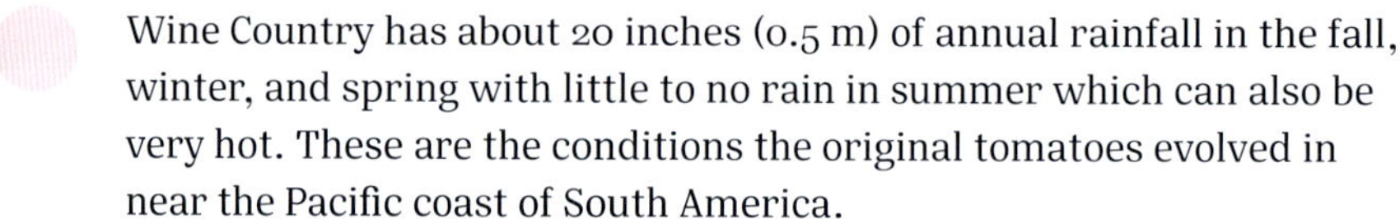

Wine Country has about 20 inches (0.5 m) of annual rainfall in the fall, winter, and spring with little to no rain in summer which can also be very hot. These are the conditions the original tomatoes evolved in near the Pacific coast of South America.

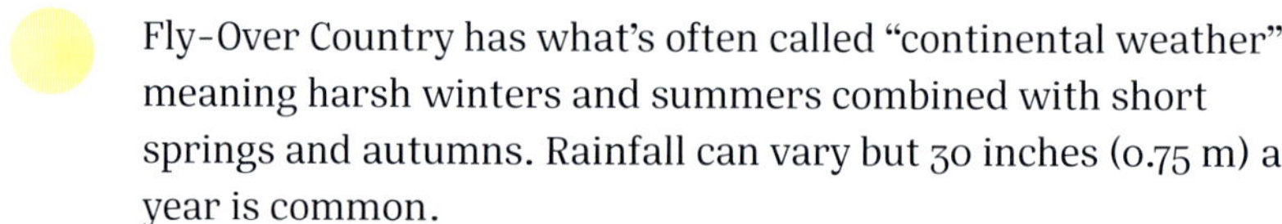

Fly-Over Country has what's often called "continental weather" meaning harsh winters and summers combined with short springs and autumns. Rainfall can vary but 30 inches (0.75 m) a year is common.

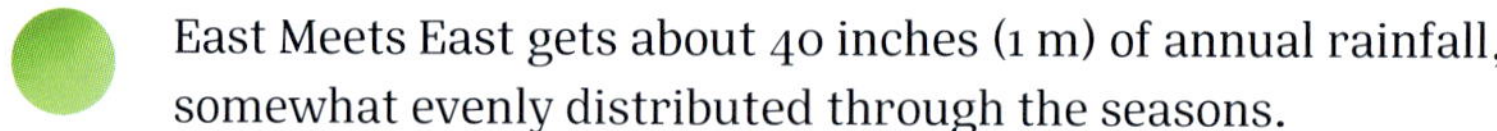

East Meets East gets about 40 inches (1 m) of annual rainfall, somewhat evenly distributed through the seasons.

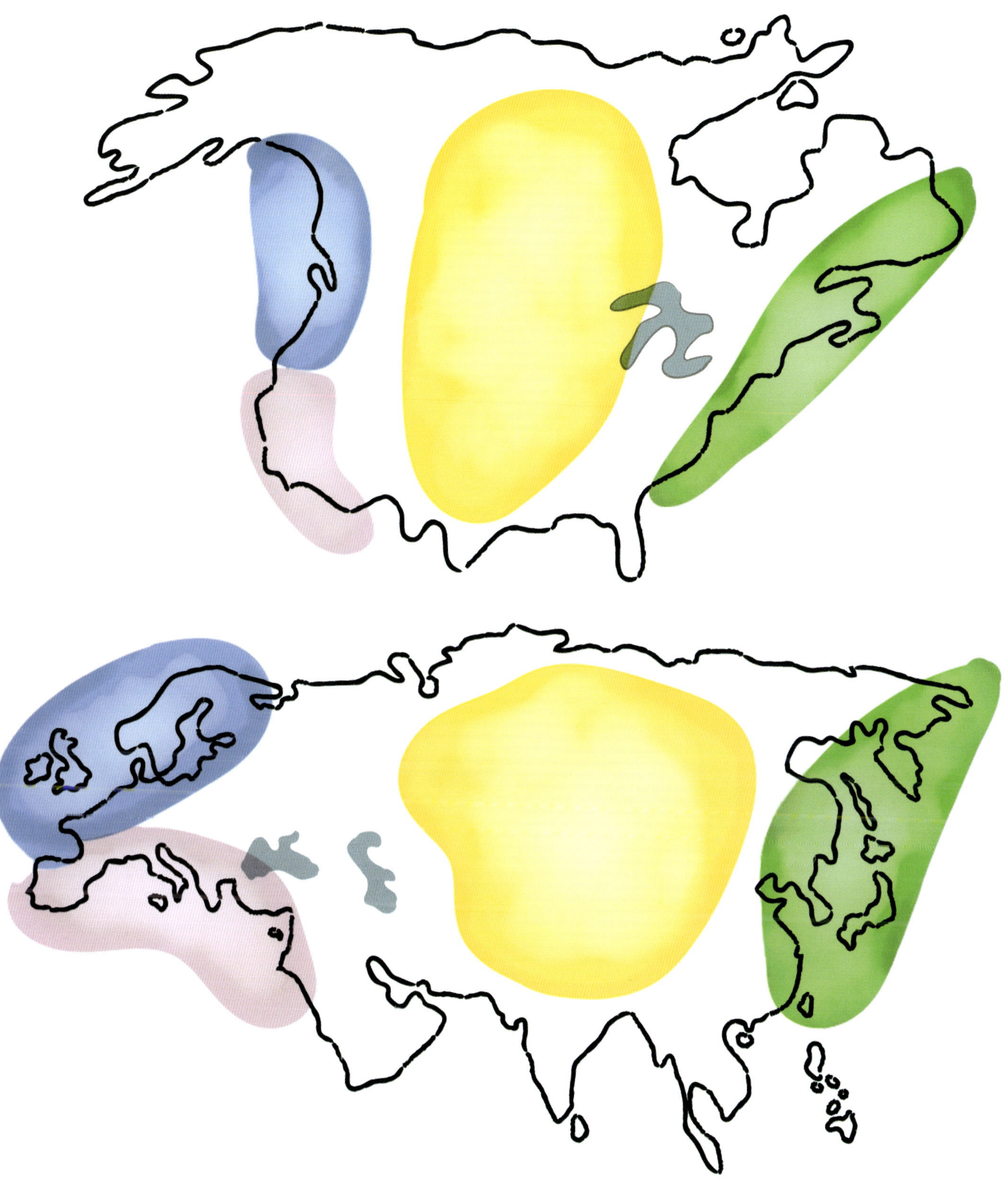

Boost Your *Weather* Wisdom

We gardeners have no control over the weather, but weather is the biggest influence on our tomato plants and the rest of the garden.

- The temperature of the soil and air determine the speed of a plant's metabolism.
- The amount of sunlight determines how much sugar a plant can produce through photosynthesis.
- Too much, too little, or just the right amount of water from rainfall determines how quickly or slowly a plant can grow. It will also determine how quickly weeds, bugs, and diseases will spread.

WEATHER ISSUES TO TRACK

- Protecting irrigation timers from freezing and cracking
- Keeping rain gauges from freezing and cracking
- Starting tomato seeds in time to plant *after* the average last frost date
- Having row cover available to protect seedlings from late frosts
- Using mulch to keep the sun from drying out the soil too fast
- Using irrigation to make up for a lack of water during dry spells
- Considering whether your region is wet enough to justify growing tomatoes under cover

LET'S *do a* DEEPER DIVE

For more details on the sister regions concept that my mentor, the late Dr. J. C. Raulston, taught me, check out this link in *American Gardener* magazine from the American Horticultural Society. There, you'll find my more detailed article titled "Latitude Adjustment" on page 30. Plant hunters from each region weigh in with their stories.

https://ahsgardening.org/wp-content/uploads/2025/03/2012-01r.pdf

The sister regions concept is helpful, but fortunately, it isn't a law. This rosemary evolved in the Mediterranean, but its "climate envelope" means it has the wherewithal to survive in the East Meets East regions, despite being pelted with twice as much rainfall as the area in which it evolved.

CHAPTER 9

Mulches: Birth Control for Weeds

LOS ANGELES GARDEN DESIGNER SHIRLEY BOVSHOW DESCRIBES MULCH AS "*BIRTH CONTROL FOR WEEDS.*" BUT, JUST LIKE BIRTH CONTROL FOR PEOPLE, YOU HAVE TO BE CONSISTENT IN USING IT. DON'T LET THE JOYS OF GARDENING MAKE YOU NEGLECTFUL OF MULCHING. IT ONLY LEADS TO HAVING MORE BABY WEEDS THAN YOU CAN COUNT.

First Lines of Defense in Weed Management

Mulch is a barrier-method of weed control. It blocks the sunlight that most annual weed seeds—like chickweed, annual fescue, henbit, and sourgrass—need for germination.

In temperate regions, a 5-inch (12.7 cm) layer of wheat straw or a 1- or 2-inch (2.54 or 5 cm) layer of organic mulch, like shredded tree leaves, grass clippings, seaweed, pine needles (rake them off at the end of the season as they don't decompose easily), or compost (if you really have a lot to spare), should last at least a growing season in an annual vegetable bed before microbes turn these dead plant cells back into water and carbon dioxide. A living mulch of low-growing annuals, like pansies and portulaca, can also work. You may also be able to buy locally available mulches that would otherwise go to a landfill, like coffee chaff, pine bark mini-nuggets, cocoa bean mulch, peanut hulls, other nut hulls, etc. If those are available in bulk—loaded on your truck or delivered via dump truck—you'll be ordering them by the cubic yard, which equals 27 cubic feet (0.76 cubic m). If they come in bags, each bag will hold 1 to 3 cubic feet (0.02 to 0.08 cubic m; see Do the Math sidebar on p. 171).

Of course, weed seeds that land on top of your mulch germinate and grow like crazy. Nonetheless, consistent mulching will cut annual weeds by 80 to 90 percent. Before digging more deeply into mulch, let's quickly cover other compatible garden practices for weed control.

WEED CONTROL POINTERS THAT AREN'T MULCH

Optimal soil conditions for your garden plants will put them in a better position to compete with weeds. A soil test will help you nail the best level of pH, organic matter, and nutrients. Find more details in chapter 3.

Soaker hoses and drip lines only deliver water to garden plants and leave the spaces between them—that weeds would seek to occupy—drier than when using a sprinkler.

Denial of light is a good way to kill perennial weeds before you plant a new bed. Since sunlight is food for plants, covering weeds with overlapping layers of cardboard or a big plastic tarp for about eight weeks will starve them out, leaving a weed-free bed. Creeping plants like Bermuda grass and vines are harder to kill this way because they can spread to find the light.

Hand pulling weeds can be character building and satisfying. For a limited amount of time. Then, it becomes a chore that sours a person's attitude about gardening. Especially novices. Do hand weed right away anything that is setting seed. And be sure to get the roots or you'll be wrestling the same plant again in a week or two. Roots come out more easily after a rain has softened the ground.

Corn gluten meal applied regularly can keep weed seeds from germinating. In hot-weather regions, you'll have to apply it more often as the heat breaks it down.

Organic herbicides such as horticultural vinegar, citrus oil, steam, and even flame can kill annual weeds by destroying their leaves, but they rarely damage the roots of perennial weeds. They may be more hazardous than some chemical herbicides for the person applying them.

Lightweight and aromatic coffee chaff—the papery layer that surrounds a coffee bean—makes a wonderful layer of mulch in the vegetable garden. And the burlap sacks that the beans come in make a lightweight layer of mulch for the pathways.

The Four Virtues of Mulch

Weed control is enough reason to employ mulch, but it has other terrific virtues. Here are four more metaphors for the many roles that organic mulches play in the tomato garden:

1. **Leaving bare soil is the number-one cardinal sin of gardening.** Not only does this damage the soil and plants, but it's a pain in the eye. You can also think of mulch as *makeup for the garden.* Like lip gloss, don't be afraid to try different ones. Using a variety of mulches helps give plant beds their identities. Depending on availability and how I'm feeling each season, we use coffee chaff or pine bark or pine needles as mulch in the vegetable garden.

Do the *Math*

How much mulch should you buy to cover a 4 × 8-foot (1.21 × 2.43 m) bed with a 2-inch (5 cm) layer of mulch?

A bed that size covers 32 square feet (4 feet × 8 feet = 32 square feet, or 1.21 m × 2.43 m = 2.88 sq. m . . . 2.9 sq. m rounded).

You don't want mulch 1 foot (30.4 cm) deep. You only want it 2 inches (5 cm) deep. And 2 inches (5 cm) equals 1/6 foot (0.05 m). So, divide the theoretical 32 cubic feet (0.9 cubic m), by 6 and you'll see that you need about 5 cubic feet (0.14 cubic m) of mulch to cover a 4 × 8-foot (1.21 × 2.43 m) bed with 2 inches (5 cm) of mulch: 32 cubic feet × 1/6 = 5.33 cubic feet. Two or three bags of mulch would cover that well enough.

Another way of looking at it: If you had three beds that were 4 foot × 12 foot = 48 square feet × 3 beds = 144 square feet.

Pretend you need the mulch 1 foot thick = 144 cubic feet (4 cubic m): 144 cubic feet × 1/6 = 24 cubic feet (0.67 cubic m). If you can get a load delivered, ask for 1 cubic yard, which equals 27 cubic feet (0.76 cubic m). Better to have a little bit extra.

2 **Any mulch also acts as a *shock absorber for rain.*** If you have bare soil in your garden—you sinner—a little poking around with a trowel will reveal a couple of things. First, the bare soil is more compacted than soil under mulch. Pounding rain does that. Raindrops may not seem heavy, but rainfall compresses the ground. This compaction harms roots by denying them oxygen.

We get about 35 inches (88.9 cm) of rain a year (YMMV!—your mileage may vary), so 1 square foot (0.09 sq. m) of bare soil is pounded by about 22 gallons (83.27 L) of water annually, which weighs about 175 pounds (79.37 kg), but that weight is amplified by the distance it falls. Mulch (and plant leaves) happily absorb the shock of that water torture so the soil doesn't have to.

3 **Your poking around in the bare soil will also reveal that it is drier than soil under mulch because mulch acts as *shade cloth.*** Sunshine sucks the moisture out of the soil. If you'd like to spend less time and money watering your garden, get religious about mulching.

4 **Last in our medley of metaphors, think of organic mulches as a *smorgasbord for soil creatures.*** That's more than a metaphor; it's a scientific fact. Digesting organic matter is the main job of the "billions and billions" (God rest Carl Sagan's soul) of bacteria, fungi, earthworms, springtails, and other creatures in a cubic foot of soil. This matters to gardeners because all this microscopic munching creates the moist, airy, nutrient-rich soil that plant roots—especially tomatoes—thrive in.

This cold-weather-loving collard plant goes into the garden just before or after the tomato plants come out in late summer or early fall. Here, the soil support system gets covered with a layer of tree leaves shredded by a mulching mower.

How to Employ a Killing Mulch on Tough Weeds *Before* Starting a Tomato Garden

Sometimes, picking out a sunny spot for starting a tomato garden means that hot-weather lawn grasses and other tough perennial weeds have already colonized that area. Some of the worst perennial weeds are grasses like Bermuda, St. Augustine, centipede, Zoysia, and others. Grasses' narrow leaves aren't the only way they are different from the broadleaf plants in our gardens. Grasses all evolved with a metabolism that allows them to grow faster than broadleaf plants, even when moisture and nutrients are in short supply; when nutrients and water are plentiful, they really grow and spread quickly. By comparison, it's like most broadleaf plants have the metabolism of a family van and grasses have the metabolism of a sports car.

But it's not just about their metabolism. It's their manner of growing: The roots that grasses and perennial weeds can spread underground can come up inside raised beds and climb to the surface to sprout leaves; their leafy structures can grow over the top of the mulch and establish new roots. They are very aggravating weeds.

You might think you can eliminate them by pulling or digging or tilling them up. But, no. Any small part of the plant that gets left behind (including miniscule root pieces) can establish a new plant that soon goes into sports car mode, defeating all your efforts.

Fortunately, these plants can be killed with much less effort than that. You just have to deny them sunlight for eight weeks during the summer growing season *before* the summer you plan to plant tomatoes. How do you do that? Well, let me tell you.

Gather enough big sheets of cardboard to cover the grassy/weedy area. Some stores will let you plunder their cardboard from their backdoor or their recycling dumpster. Let each sheet of cardboard overlap a neighboring sheet by about 10 percent or so. Then, to keep the summer grass or weeds from snaking their way out between the sheets, cover all the cardboard with at least a couple inches of wood chips—often they can be acquired for free from arborists—or any other plentiful mulch. The weight of the mulch will seal the gaps where the cardboard sheets overlap, entombing the grasses in a very dark place.

No plant can survive being deprived of sunlight for eight weeks: It's their food, after all.

Key caveat: *This only works during the grass's growing season.* If you try it near the end of summer, the grass is already getting ready to go dormant for winter. The darkness of the cardboard will just seem like an early winter and they'll go dormant sooner. If you try it in fall or winter or early spring, the grass is dormant and can survive without sunlight. The cardboard will have decomposed in a few months and those grasses can "wake up" and punch through your soft, decomposed mulch. All your effort will have been wasted. But if you cover the grasses during their growing season, they will die and you can start your tomato garden the following year with a fresh, clean slate.

A Living Mulch of Edible Flowering Plants

These paste tomatoes with a cat's cradle trellis are mulched by annual pansies through the early, cooler season and by annual portulacas in the summer. Both plants are edible and delicious. I prefer yellow flowers for their welcoming color.

Can you grow pansies in fall or spring before tomato season starts? Can you grow portulaca or purslane during summer's growing season? If so, you could have a vibrant, brightly colored living mulch all through your growing season. Plus, these plants are edible and delicious.

Pansies (*Viola × wittrockiana*) are cool-season annuals. They can be planted in your beds well before the tomatoes, or even at the same time. If you have hot summers, they'll fade out in the heat. Planted 8 to 12 inches (20.32 to 30.48 cm) apart—with a fistful of organic fertilizer in the hole—they will make a solid mass about 4 inches (10.16 cm) high and will block almost all your weeds by denying light to the weed seeds. Every part of the plant above ground—stems, leaves, flowers, buds—can be clipped or pulled off easily. Pansies are edible and go well with salads, sandwiches, and snacking. Emphasis on "snacking" because you don't want to thin them to the point they aren't acting as an attractive mulch. I plant them partly to block sunlight from drying out the soil and partly just to make the garden bed beautiful and partly to enjoy snacking on them in the garden or embellishing a dish in the kitchen.

Other similar plants have the same qualities and can be used in the same way: Johnny jump-ups, violas, and violets are all smaller but similar edible cool-season plants that could play the same role.

Portulaca (*Portulaca grandiflora*) is a South American plant that's also called moss rose. Plant them about 8 inches (20.32 cm) apart with a fistful of organic fertilizer in the hole to create a living ground-hugging mulch around your tomato plants. All parts of the plant above ground—stems, leaves, buds, flowers—are edible. They're more succulent than pansies, but still great for salads and sandwiches in the kitchen and a little light snacking in the garden. Portulaca does not like frost so it needs to be planted after the weather has warmed. The plants will last until the first or second frost of fall.

Coffee *in the* Garden

Having coffee in the garden can mean one of three things:

1. Coffee for drinking
2. Coffee for composting
3. Coffee for mulching

FOR DRINKING

My wife and I believe that half our time in the garden should be spent unwinding rather than working. We can be found enjoying hot coffee in the morning or iced coffee in the afternoon while reading the paper or a novel. It's one of the best ways to relax while enjoying the play of the breeze and the antics of the birds. Top it all off with a dose of vitamin D and the satisfaction of looking at your growing garden, perhaps from within your straw bale and tomato snacking courtyard.

FOR COMPOSTING

Perhaps 5 percent of what powers our compost comes from our kitchen's coffee grounds and coffee filters (the caffeine doesn't affect the compost but could hurt roots if used as mulch). Because coffee grounds are high in nitrogen—like other kitchen scraps—they are considered part of the "green" portion (see p. 185) of a compost concoction. Another 10 percent of what powers our compost comes from dry, fluffy, organic coffee chaff—as in "separating the wheat from the chaff." We use that for the "brown" part (see p. 185) of the compost to cover the "green" portions. We store the chaff for

When Does Weeding By Hand Make Sense?

A good layer of mulch is valuable for many reasons, but it's not 100 percent protection from weeds. Airborne annual weed seeds can land on top of the mulch and germinate. Many perennial weeds can tunnel their way underground or crawl their way above ground to invade your tomato beds. You'll have to stay on guard with the weeds just as you will with bugs and diseases.

Fortunately, in a tomato garden, it's not necessary to have 100 percent elimination of weeds throughout the season. For most annual crops, like tomatoes, if you can keep the beds clear of weeds through the first two-thirds-ish of the growing season, your tomatoes will have enough time and space to establish a robust root system that will give you plenty to harvest.

composting in cast-off garbage cans—galvanized or plastic—to keep it dry and next to the compost bins. More details about chaff follow.

Then, before vegetable planting in spring and fall, you can find us spreading coffee-powered compost in the vegetable beds. See pages 186–192 for details on our "no turning and no stinking" compost system.

FOR MULCHING

Coffee chaff is the beige, papery layer that comes off the green coffee beans when they get roasted to a gorgeous brown color. Our local, organic coffee roaster bags up their light, dry, fluffy chaff and leaves it in a covered shed for gardeners to pick up for free. Otherwise it's just a waste product they would have to pay to send to the dump, unless they could find a company to compost it on an industrial scale.

A full bag of chaff is shockingly light in weight. Almost any person can lift one full bag in each hand and toss them easily into the back of their vehicle. One big plastic garbage bag of chaff will cover a 4 × 10-foot (1.21 × 3 m) raised bed with about 3 inches (7.62 cm) of mulch. After a few rains, the chaff will be half that thick, but still plenty thick enough to hold moisture in the ground and to suppress annual weeds. By next year, you can easily turn it over or just plant through it.

Bonus: The garden will smell like coffee for a few days.

If weeds are spreading during the last third of the season, the modest amount of crop loss—from competition for water and nutrients—will often be less than the amount of effort it would take to be 100 percent weed-free for that last stretch.

The bottom line is that if you don't have the resources to keep weeds down for the whole season, put your best efforts into the beginning of the season. You want to give tomato roots the best chance to spread and grow without competition.

The exceptions to this general rule about weed competition are if your very short micro dwarf varieties are getting shaded out by tall weeds, like common ragweed (*Ambrosia artemisiifolia*) or foxtail grass (*Setaria* sp.), or if your tall tomatoes are being shaded out by climbing vines like morning glories (*Ipomoea* sp.). Tomatoes can tolerate some late-season competition for water, but they have zero tolerance for losing out on sunlight, which is their food supply. Late-season weed shade will slow ripening and reduce fruit size and numbers.

WILL YANKING PERENNIAL WEEDS WORK?

Not really. Hand weeding perennial weeds in the tomato garden can make the garden look better for a few days or a week, but the weeds will come back and continue their vigorous root competition unless you get out *all* of the root, which is not easily done.

Perennial grasses, like Bermuda, St. Augustine, centipede, Zoysia, etc., not only have the most vigorous, competitive metabolism of all plants, as previously noted, but they have deep, grippy roots that don't come out easily, or at all, with hand weeding. Using a hand tool—trowel, hoe, any kind of "weed digger"—can give the impression of success. But, any scrap of perennial grass root left behind will revive like a zombie in a movie. It won't eat anyone's brains, but it/they will compete with tomato roots for water and nutrients in short order.

There are annual grasses, like crabgrass (*Digitaria sanguinalis*), that have that speedy metabolism too, but their roots aren't as feisty as perennial grass roots. So, unlike their big brother and sister grasses, they can be controlled by hand pulling.

Perennial weeds with bulbs or tubers, like nutsedge (*Cyperus* sp.), give the impression of being easy to control, even older plants can be pulled from moist ground and give gardeners the temporary satisfaction of seeing what looks like its root system come out of the ground. But the tuber—also called a "nutlet"—is designed to be left behind to grow again if the top is pulled or eaten. There's a whole extensive root system under that tuber that continues the competition game.

Perennial weeds of any species are best controlled during summer before you start your tomato garden.

Is It *Really* a Grass, or Does It Just *Look Like* One?

Botanists have a popular old saying that helps even a novice differentiate between grasses and their—sort of—look-alikes: "Rushes are round, sedges have edges, and grasses have hairs."

All three types of these plants might be in your garden or neighborhood. The leaves of rushes are tubular on the inside and round on the outside like a straw. They like ground that is too wet for growing tomatoes, so you might only see them near bodies of water.

Sedges—like the nutsedge mentioned previously—have edges, meaning that if you put your finger and thumb around the upright leaves, you'll feel that, together, they have a triangular profile.

You can tell if a plant is a grass by looking for joints between the leaves and the stems. At that joint you'll see some short, stiff hairs.

Sourgrass is a common annual summer weed that vaguely resembles clover. The leaves—right—which resemble a Valentine heart, have a sour, lemony taste. The seed pods—left—have a similar flavor, but they also pop in your mouth.

WILL YANKING ANNUAL WEEDS WORK?

Yes! Annual weeds in the tomato garden can be managed easily by hand weeding. Your common annual weeds like sourgrass (*Oxalis acetosella*), purslane (*Portulaca oleracea*), and prostrate spurge (*Euphorbia maculata*) all arrive as airborne seeds that land on top of your wonderful mulch. They all have tiny seeds, which is nature's way of telling you they need sunlight to germinate (larger seeds, like corn and peas, tell you they need to be in darkness under the soil, at a depth of about three times their diameter to germinate). All the annual seeds under your mulch will leave you be, because they can't germinate in darkness.

The good news is that annual seeds germinating on top of your mulch will have shallow, weak roots that can be pulled out easily by hand, especially after a rain, when the mulch is soft and yielding. The smaller they are when you pull them, the more easily they will come out.

The Buzz on *Weed-Free* Garden Paths

I like to have lots of gardens and lots of things growing, but I don't want to find myself forced to spend too much time or too much money taking care of my garden. Sound familiar? Managing weeds is one of the biggest potential time and money drains, especially in a vegetable garden. And garden pathways—the spaces between beds—can be one of the most challenging areas to weed, especially if you want your garden to look good too. Free wood chips for mulch are one option, but it's a lot of labor to haul and spread them every single year. Gravel is more permanent but it won't stop many weeds. You could allow grass to grow in your paths, but then you've got a weekly mowing and clipping chore. Bricks for paths? That's expensive. And even if the bricks are free, it's a big job to install them and weeds will pop up in the cracks.

For years I've looked for a better answer. Fortunately for us, we have an organic coffee roaster in town and I depend on them for more than my morning java and the coffee chaff and grounds I use in my compost pile. The green coffee beans come in burlap sacks. Once empty, the employees stash the bags behind their building under a little shed roof before tossing them in the dumpster. So, I swing by there twice a year to throw a bunch of bags in my truck. Then, I overlap them in our pathways. You want to get bags from an organic roaster so you'll know you aren't adding pesticide residue to your garden.

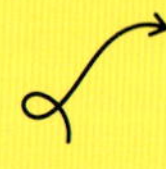

A classic Italian Caprese salad composed of any tomato, basil leaves, and fresh mozzarella cheese.

It only takes ten minutes to load the truck and another fifteen minutes or so to lay the bags over about 150 linear feet (45.72 m) of paths. I'd easily spend six times as much time spreading wood chips in an area that size *if* I could get them dumped close by.

I lay the sacks with about a 50 percent overlap so the entire path is covered with a double layer of burlap sacks. They have enough substance so they don't blow away like cardboard. Keeping the soil underneath in darkness eliminates about 95 percent of the weeds by suppressing germination.

But, some weed seeds will germinate on top of the bags. This is the fun part: flipping the bags after two to three months pulls up the few remaining weeds and exposes their roots to the sun so they dry out and *die*. In another three months, the bags have decomposed to the point where it's time for another layer and another trip to the coffee roaster. On days when my trip coincides with their free coffee tastings, I get the work done faster.

LET'S *do a* DEEPER DIVE

A surprisingly large number of weeds are edible, so you can do double duty: remove weeds from the garden and add free food to your salads, sandwiches, and side dishes. These three books can help you identify common weeds as well as teach you which ones to set aside for your table (or at least for your backyard hens!).

Backyard Foraging by Ellen Zachos

Good Weed Bad Weed by Nancy Gift

The Scout's Guide to Wild Edibles by Mike Krebill

CHAPTER 10

Best-in-the-Nation Compost Station

THERE ARE TWO KINDS OF COMPOSTERS—WHICH ONE ARE YOU?

Casual Composter	Aggressive Composter
◆ Only need finished compost for spring and fall planting seasons	◆ Want to turn kitchen and garden scraps into finished compost in a matter of weeks, because you need compost many times a year
◆ Don't want to spend a lot of unnecessary money on bins and supplies	◆ Have a big budget for bins and supplies
◆ Don't want to spend a lot of unnecessary time on turning and watering the compost as long as it never smells bad	◆ Have the time and inclination to turn piles by hand and don't have competing activities calling for your time and energy

Compost From the Inside Out (From Inside the House Anyway)

The word "compost" comes from Latin for composite. And that's how it starts. You compose a composite of food scraps that becomes compost. Got it?

Question: Can Meat and Dairy Be Composted?

Answer: Technically, yes, but because of smells, slower decomposition, potential for *E.coli*, and their greater appeal to vermin, I recommend leaving those ingredients to municipal composting (if that's offered) or to aficionado composters who'll make the extra effort. If that's not you, take the easy route—as I do—and give meat and dairy scraps to the backyard hens—yours or a neighbor's. They are omnivores; they value meat, fat, seafood, and dairy over other foods and will make quick work of it, turning it into eggs and manure. Share those scraps in the morning only so the hens have time to polish them off before sunset, because nighttime is when vermin clock in to work.

From inside your house—mostly the kitchen—your composite can start as anything that used to be a live plant. For example:

- Scraps of raw or cooked vegetables and fruits, including skins and seeds
- Any edible mushroom scraps
- Every kind of grain, such as stale or moldy bread and crackers, leftover rice or pasta, weevil-infested flour, etc.
- Any kind of stale or moldy nuts and their shells
- Soft paper products, such as used napkins and paper towels; some people shred paper and cardboard, but those take too long to decompose so I just recycle them. Ditto for "compostable bags"—they also take too long and are fine to recycle with paper.
- Random kitchen scraps, such as spent coffee grounds including the paper filters; whatever is in the sink drain; tea bags, including the tea bag's string, paper label, and even the staple, which will rust away to nothing, so needn't be removed.

Compost from the Outside In (from Outside the House Anyway)

Everything outside your house—meaning scraps from your yard or garden *that used to be alive*, can be fair game for your compost bin. For example:

- Grass clippings in moderation. The truly best place for grass clippings is to clip them with a mulching mower and let them dry up in place and sift down to feed the soil that supports your lawn. Easy-peasy. But one or two mower bags of lawn clippings per compost bin isn't too much, especially if you let it turn brown and use it as the brown ingredient (see sidebar).
- Soft vegetable garden and herb garden scraps that aren't thick like potato and tomato vines or the woody parts of rosemary and lavender
- Green moss and even Spanish moss
- Fallen leaves, shredded or otherwise, but not the big brown paper leaf bags. They take too long to break down. Let someone reuse them. Pine needles are too thick to decompose in a few months. They work better as an attractive mulch in perennial beds.

- Perennial weeds, roots and all—*if* you leave them exposed in the sun to die first, and *if* you don't include any of their seeds. Live perennial roots and mature perennial seeds can survive in most compost bins and wreak havoc.
- Annual weeds, roots and all—they won't survive in the darkness of a compost bin

However, some things are too big or too thick to compost in a few months, such as fallen branches and clippings from hedges and ornamental grasses. All of those, when dry, are good candidates for the firepit. Otherwise, wet or dry, they can go to municipal yard waste collection, if you have that option. If you have a wooded yard, spread them and sprinkle brown soil over them to speed up natural decomposition.

When Is a Compost Pile Like a Soil Salad?

Composting practices described in books often resemble one part sweaty gym class and one part rocket science. Combining the strict ratios of carbon to nitrogen and the expectation that you'll be turning damp, heavy compost twice a week, you have to wonder if you're going to get a headache or a backache first.

But, composting can involve much less work and bother. Every spring and fall we get a bin of five-star compost without *ever* turning it. Plus, it *never* stinks and doesn't draw *vermin*. You can do this too, and here's how.

You don't have to be Julia Child to feed your family and you don't have to be Mr. Green Jeans to feed your garden. If you've ever made a dinner salad, you have all the skills necessary to create a winning compost salad. All a high-functioning compost pile needs are the same components as a dinner salad:

- The right container
- Leafy stuff
- Some chunky vegetable stuff
- A splash of dressing

Brown and Green Ingredients

"Brown" refers to common compost ingredients that are low in moisture and low in nitrogen, like fallen tree leaves—shredded or whole—dried grass clippings, coffee chaff, wheat straw, and other waste products that are usually pretty dry and brownish in color.

"Green" refers to common compost ingredients such as kitchen scraps, garden clippings, pet hair, fresh grass clippings, and other waste products that have moisture and nitrogen.

Some compost guides focus too much on getting some precise ratio of brown to green. My simple system is to add green components to the compost bin first and then apply enough brown components to cover the greens so you can't see them. That's enough brown to keep away flying insects that would infest the greens.

The microbes—in your small pot of last year's compost—that decompose this year's browns and greens into finished compost will work well as long as they have access to both materials. They're not calculating ratios.

But, for the purposes of our discussion about *compost* salad—because it's similar to, but not the same as, a *dinner* salad—I'm going to switch up the *order* just a bit and add details about seasonal timing. Don't worry, it will all make sense as you read through these steps for the components of a compost salad:

- The right container
- Some chunky vegetable stuff
- A splash of "dressing"
- Some leafy stuff
- Seasonal timing

THE RIGHT CONTAINER

Start with two compost bins sturdy enough to deter urban wildlife (I'll address the need for two bins in the Seasonal Timing section). The best bins have a hinged lid so you can open them with one hand while holding ingredients with the other hand. If your two bins don't come with slots to allow rain in—because a dry compost is an inactive compost—drill a dozen ½-inch (1.25 cm) holes in the lid. During dry spells, hose down the compost really well.

Place the bins where they'll be quickly accessible from the kitchen door. If you have chickens, you could instead place them next to the coop so you can sweep old bedding directly into the compost bins. If you're in a colder region, put the bins where they'll get sun in winter to keep the composting bacteria warm enough to do their work. In warmer climates, they can be in shade all year and will still finish composting on schedule.

Our compost station before we had chickens. Clockwise from bottom left: full bin that's almost composted, empty bin that's ready to reload, large nursery pot with shredded leaves (browns), kitchen scraps (greens) in square container, garden soil/microbe source in nursery pot.

Compost station moved behind the coop so bedding and chicken manure can be swept directly into the bins.

Scraps of rubber pond liner make it easy to sweep coop waste into the bins.

To eliminate low spots that vermin can enter through, use scrap wood and/ or bricks to create a flat seal between the bottom of the bin and the hardware cloth.

The bottom of the bin is important too. You don't want vermin using the bottom as a crawl-through fast-food joint. Make sure the area under the bin is flat enough that a sheet of metal hardware cloth on the ground—with the same footprint as the bin—will keep vermin from tunneling into the bin. Make sure the bottom of the bin meets the hardware cloth all the way around on all four edges. You may want to make this protection sturdier by sandwiching the hardware cloth between the bottom of the bin and a line of bricks. Alternatively, you could make a flat floor of bricks to cover the entire bottom of the bin, instead of using hardware cloth. Just be sure not to leave any uneven places where a rat can slither in and eat up kitchen scraps and multiply.

If your bin doesn't have doors that open near the bottom, don't fret. Those doors are a poor idea: that you can drop raw kitchen scraps into the top while shoveling finished compost out the bottom. Wrong. Ignore those doors. They rarely open or close easily anyway. When each bin has finished composting (see Seasonal Timing) scoop out what can be easily reached with an open lid. Then, tilt the bin back and forth until you can lift it or tilt it on its side, exposing the bottom of the pile of finished compost to be scooped up off the ground easily with the shovel. Then, double-check to make sure your hardware cloth and/or brick layer is properly in place and keeping the bin vermin-free.

HOW TO MAKE A CRITTER-PROOF COMPOST BIN

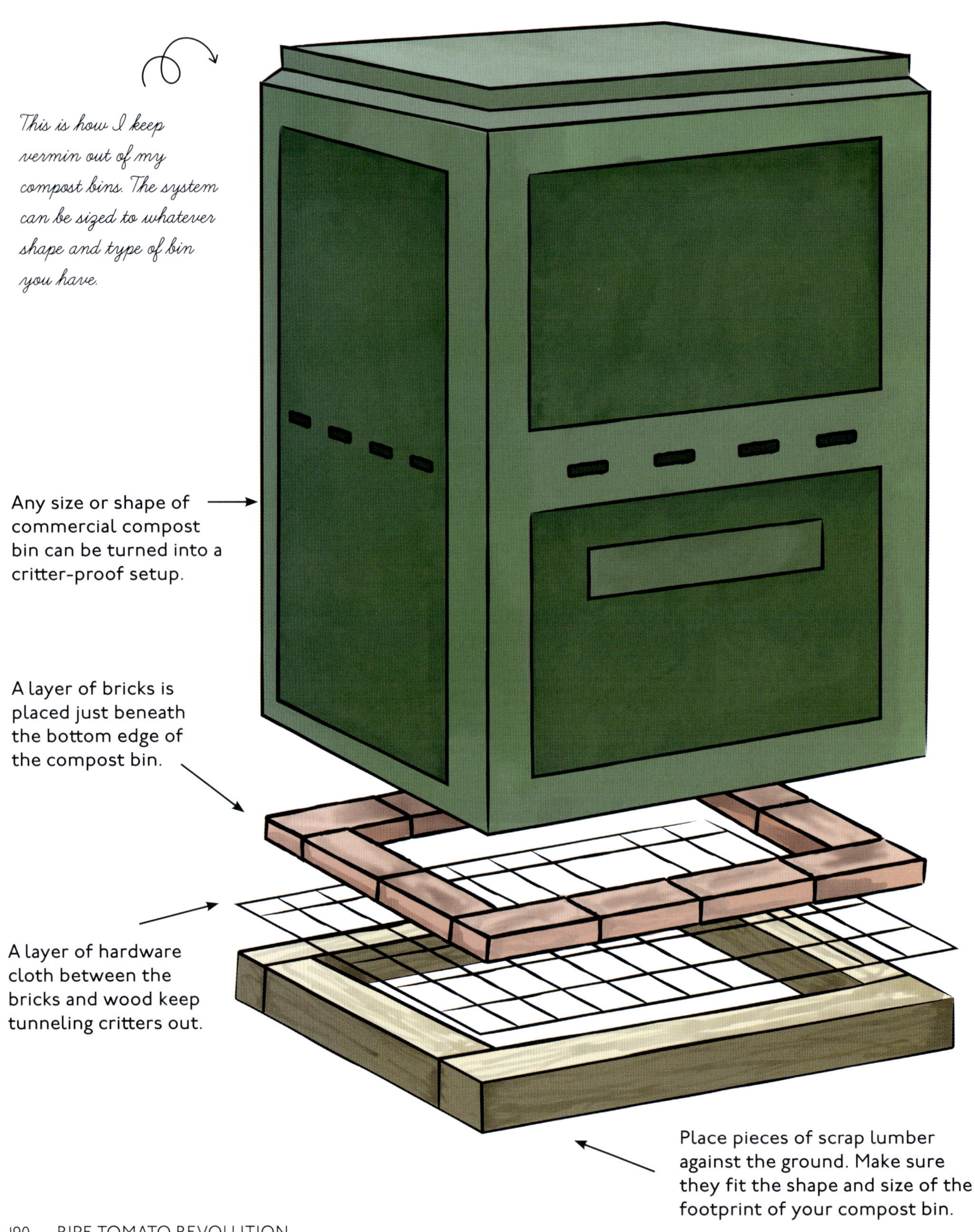

CHUNKY VEGETABLE STUFF

To start your compost salad, throw in your chunky kitchen scraps or garden scraps like dead plants, pruned leaves, and spent flowers. Just as with a tossed salad, we don't want to overwhelm the leafy part with too many chunks. A tossed salad with too many chunks looks like an uncooked casserole, and compost with too many chunks ends up too soggy and dense and could start smelling bad. Yuck. Think basic salad, nothing fancy. Let only about half of what's in the pile be chunky scraps, and compose your yard salad the same as you would a tossed salad for a vegan friend—no meats, cheeses, or grease. Those items tend to gum up the works and take too long to compost.

We save our kitchen scraps in an airtight container that stays on the counter. I know people who also store their scraps in the fridge if they only generate small quantities that would sit around too long waiting for a trip to the compost bin. You can toss vegetable scraps, stale bread, eggshells, paper towels, coffee grounds, coffee filters, and old flower bouquets in your compost.

Scraps from the garden can also go into the compost, but unless you want problems, don't include seed heads from weeds (feed those to the chickens!), diseased plants, or anything recently exposed to herbicides or pesticides. The chemicals could kill the good microbes and plant diseases could be spread. The weed seeds would just get recycled back into the garden, unless you work extra hard to ensure your compost pile gets hot enough to make steam. If that's the case, take a piece of cardboard and cut yourself a personal "Real Hot Composter" badge and maybe you can write the next book on composting!

Chicken Salad?

If you have chickens, you may want to do as we do and keep two scrap containers on the counter: one for chunky scraps suitable for chickens like eggshells, vegetable peelings, stale bread, leftovers, etc., and another for the compost bins that includes spent coffee grounds, tea bags, paper towels, banana peels, citrus (chickens don't like it), avocado (can poison chickens), etc. Sweeping the manure-infused spent bedding from the coop floor twice a year into the compost bins will create some excellent finished compost for your tomato beds and containers.

Olga tosses hen-friendly kitchen scraps to our chickens so they can create eggs for us and fertilizer for the tomato garden.

A SPLASH OF "DRESSING"

Sprinkle a bare trowelful of garden soil over the kitchen scraps. This is important. This garden soil "dressing" distributes *composting* bacteria throughout the pile—from bottom to top as it builds up over time. These microbes are where compost's earthy aroma comes from. Composting bacteria will outcompete the *rotting* bacteria (that is on all food items) that cause kitchen scraps to stink when they aren't properly composted. Different bacteria species mean different smells.

Even a thin layer of dressing provides the punch that makes a dinner salad great. Without any dressing, a salad is just health food that gets skipped at the potluck. A compost pile without its metaphorical dressing will get passed over by the garden gods as well. Just as we have choices for salad dressings, we have choices for compost dressings. Some of last year's compost would be ideal. You could also use brown soil from a neighbor's garden if you're just getting started, or soil from a woods or meadow.

If the soil is brown, it's because it contains *organic matter*, and if it has organic matter, it also has *composting bacteria*. And you can get this for free. Anyone selling you "compost activator" doesn't understand composting and is just separating you from your money.

You won't need much of this dressing. Fill a 1-gallon (3.78 L) nursery pot with garden soil (or last year's compost), stick an old trowel in it, and stand it on the compost bin that's *not* getting filled up.

A nursery pot won't hold water and will drain when it's in the rain, and the bacteria will survive whether it's wet or dry. The trowel we use is one that's missing it's handle. So, reuse!

By spreading this tiny bit of dressing on each dose of kitchen and garden scraps, you're doing the *very light work* of distributing the composting bacteria throughout the pile, from bottom to top, without needing the *heavy work* of being turned.

When people turn compost with a pitchfork, they believe the benefit comes from the distribution of oxygen throughout the pile—which does happen a bit—but this is a misunderstanding of something happening that's more important, and it explains why those tumbling compost bins don't really work. Instead of mixing oxygen in the pile, the hard work of turning with a pitchfork merely distributes the organic matter/bacteria, which is at the very bottom of the pile, throughout the pile. But you could do that distribution instead with the light work of sprinkling brown soil "dressing" as you go. Easy-peasy.

LEAFY STUFF

And now, to bring things full circle, cover the vegetable and garden scraps and the "dressing" with brown leafy bits. This not only keeps flying bugs down, but the high-carbon leaves also balance the nitrogen in the kitchen scraps. There is no need for precision; just toss in a double fistful or so, enough to cover the chunky scraps and dressing.

The leaves you use can be either shredded or whole. The "dressing" will take care of breaking them down.

Some species, like willow oak leaves, break down better if they're shredded and bagged by a lawnmower first, but most leaves will compost fine, even when added whole. If your yard is treeless, harvest the labor of the good folks who spend lots of time and money raking and bagging their leaves and putting them on the curb for you.

Store the leaves next to your compost bins in a large, covered container, like a garbage can, to keep them dry.

SEASONAL TIMING

Hopefully, you put "compost bins" on your birthday wish list and you can start right away with your best-in-the-nation compost station.

If so, start filling your first vermin-proof bin over this winter and spring with leaves, chunks, and dressing. After it's full and before the beginning of summer, leave it alone to decompose quietly. Then, start filling the second one. By fall, your first bin will contain finished or nearly finished compost. Feed this compost salad to your garden and then start filling that bin again. The second bin will then have time to finish composting for spring planting.

Here's a sample schedule in which you are either **filling one of the two bins** or **leaving it alone** and letting it finish composting. Put your nursery pot with last season's compost—or dark garden soil—and a trowel on top of the bin that is NOT being filled, so it's easy for you or a guest to know that bin is not to be opened. Then, follow this schedule to have finished—or nearly finished—compost available from one bin every spring or fall:

- Fill the first bin for six months, then stop.
- Start filling the second bin for six months.

- When the next planting season arrives—spring or fall—empty the first bin into the garden if it's all dark and decomposed, or nearly so.
- Now that the first bin is empty, start refilling it.
- Stop filling the second bin and leave it alone until the next planting season.
- Maintain this process for every planting season: Empty one bin each spring and fall.
- And then start filling the empty one.
- Park the pot of last season's compost on top of the full bin that's being left alone to compost until next planting season.

With this compost salad process, you'll have a bin's worth of finished compost every spring and fall. Neither bin will stink, you won't attract vermin, and you'll never have to turn compost again. Ever.

With a set-up like this, we:

- Get lots of compost.
- Keep lots of material out of the landfill.
- Have a robust garden.
- Don't really spend a lot of time or energy fooling around with the compost—the only thing easier would be tossing a salad.

Two medium garbage carts parked near the kitchen door. A salvaged garbage can keeps browns—leaves and/or coffee chaff—dry. salvaged concrete HVAC pads keep the area from getting muddy.

Conventional Compost Bins and Practices That Disappoint and/or Cost Too Much

I only like one style of compost bin. It has a hinged lid that could be opened with one hand (while the other hand carried the kitchen scraps or garden weeds). The lid also had holes in the top to allow rain in, which helped keep enough moisture in the pile to sustain the composting bacteria and fungi.

But it had several problems:

- **Needlessly expensive:** It was just so expensive for a box made of plastic. Compost bins generally sell for almost $300 to over $500 each, including shipping.
- **Useless doors on the sides:** Part of the expense is due to one or more doors near the bottom of the bin with the "idea" being that while you're adding more kitchen and garden scraps through the top, you can squat down and shovel out finished compost from the bottom. Squatting is a good gardening skill; but shoveling from a squatting or sitting position is not realistic. And the idea that fresh materials are being added at the top and finished materials can be scooped out of the bottom is just unrealistic. Those doors at the bottom are superfluous and unwieldy anyway you look at it.
- **Insulation *and* ventilation?** Some advertise that they have "insulation to keep the compost hot even in winter" and—in the next sentence—describe that they have "ventilation slots to provide aeration." Uh. Nope. You can't have both insulation *and* ventilation and think you're going to contain any heat. That's like spending oodles of money on insulating your house and then leaving the windows open all winter. The temperature inside your house—or compost bin—will roughly match the temperature outside. Don't buy this hype.
- **An open bottom that attracts vermin:** The entirely open bottom requires covering with hardware cloth—like metal fencing, but with smaller openings. But, without perfectly flat soil, the hardware cloth won't keep vermin out.
- **Expensive and unnecessary compost "starters":** You can buy a bag of compost "starter," but it won't have anything that any brown soil you have access to has. If the soil in your yard or garden or last year's compost is brown, that color means it has decomposed organic matter. When soil has organic matter, it means it contains decomposing fungi and bacteria.

LET'S
do a
PROJECT

Compost on Wheels

PERHAPS, LIKE ME, YOU MIGHT THINK CONVENTIONAL COMPOST BINS:

- Are too expensive (especially since you need two bins to manage compost well)
- Are difficult to unload from the top
- Are almost impossible to unload through the lower door
- Break too easily—judging by some reviews
- Are too unwieldy to put together
- Are too accessible for rats given their wide-open bottoms
- Absolutely worst of all, don't even help you carry the finished compost to the garden!

One day, feeling certain I could figure out a better and less expensive way to compost, I went up and down the container aisles of a big-box construction supply store. Looking, checking, imagining, wondering. Finally, I had my "Eureka!" moment right there in aisle nine. Wheeled garbage cans!

I could choose the size of my composting containers, save money, and have as many of them as I wanted/needed. The solid bottoms would block vermin instantly. My new compost bins were made to stand up to every kind of weather; there was no need to put anything together; they would be soooo easy to unload; they were designed to be banged up for decades and not break; and, as an added bonus, these "garbage carts" with two wheels were *mobile*.

Once you get your wheeled trashcan home:

After putting them into action, I've learned that when a cart is ready to dump, I don't have to shovel or fork it out from an uncomfortable posture. I merely have to wheel it to the vegetable garden, flip the lid open, tip the can upside down onto one of the beds, and let the compost pour out. No tools needed! No bending! Just dump and go!

If you lack the upper-body strength to tip it upside down, you only need to hold the lid open, tilt the cart back until the lid props the cart at a 45-ish-degree angle, then shovel or fork the compost out from a comfortable standing position. Problem crushed.

All any gardener has to do is get a couple of these mobile composting cans home. No truck or SUV? Some big-box stores will deliver these for about the same as the shipping price for the more expensive compost bins.

Easy-peasy, as my carpenter girlfriend of so many years ago would have said.

Olga dumps kitchen scraps—fruits, vegetables, paper towels, coffee grounds, filters, and more—into a mobile compost cart.

EXTRA TIPS FOR SUCCESS

- After drilling out the top and bottom drainage holes and side airflow holes—and disposing of all the yucky little plastic spiral scraps created thereby—I put my weight onto the middle of each of the rounded lids so they bend down with a "pop" to make a dimple that allows rainwater to flow into the new drain holes.

- If you have trouble with raccoons—or, as we call them, "trash pandas"—you could stretch a just-the-right-length bungee cord over the top from the handle at the back to the handle at the front. Placing a brick on each lid would probably work too. Raccoons are clever but they're no Superman.

- And, one last step for the installation: Put a 1-gallon (3.78 L) nursery pot full of last year's compost or brown soil and an old—or even broken—trowel on top of the lid of the bin that's being allowed to finish composting. Be sure to leave the top of the bin that's being filled uncovered so all your household members will see right away which one is waiting to be topped up and "dressed."

To avoid the onerous and unnecessary hassle of turning a compost pile, Olga merely adds a trowel scoop of last year's compost or brown soil to inoculate new scraps with composting microbes.

Dry leaves from autumn (or coffee chaff)—stored in a water-tight container like this salvaged galvanized garbage can—add the "brown" component and reduce flying insects. Holes in the depressed lids let rainwater in to keep the microbes hydrated and busy.

Quick Reference *Dos* and *Don'ts* of Backyard Composting

Some backyard compost bins are neglected, stinky, and unproductive. Others are overmanaged to the point of demanding too much time and effort. Get your compost compound working for you with much less effort and greater success with these Dos and Don'ts.

DO:

- Recycle the garden by tossing most garden clippings into the compost bin. But save fragrant, resinous herb clippings (they need a harsh annual pruning to keep from getting leggy) in a watertight container for use as fire starters in winter. And toss weeds that have gone to seed into the hen pen or the trash; you don't want your compost to become a weed distribution center.
- Maintain a vegan-ish compost diet. Scraps from fruits, vegetables, herbs, and nuts will compost well. Eggshells, paper towels, tea bags, and coffee grounds are also okay. Forgo meats, bones, fats, oils, and dairy, including cheeses; they break down too slowly.
- Maintain domestic tranquility by gathering kitchen scraps in an airtight container that won't attract fruit flies, or store the container in the fridge or freezer.
- Keep twin bins—the first one gets fed scraps while the second one quietly digests. That way, you'll always have finished compost from one bin for spring planting and the other bin for fall planting.
- Keep an eye on things by putting compost bins close to the kitchen door—otherwise they're out of sight and out of mind. Mine is not far from the backdoor and next to the chicken coop for direct disposal of bedding and manure.

DON'T:

- Be a vermin magnet by leaving kitchen scraps in a compost bin that rats, mice, raccoons, and possums can access. When composting kitchen scraps, use heavy-duty plastic bins with tight-fitting lids and a 3 × 3-foot (91.44 × 91.44 cm) square of ½-inch (1.25 cm) hardware cloth underneath to keep out critters.
- Create a stinkpot. All food scraps come with a ready-made population of stinky, rotting bacteria. Combat this with occasional sprinkles of garden soil from an old trowel and a nursery pot (the holes will allow rainwater to drain away). Any soil that's brown has organic matter and any soil that has organic matter has an indigenous population of composting bacteria that can outcompete rotting bacteria. No need to buy expensive "compost activators" when you have brown garden soil.
- Break your back turning compost. The main benefit of turning compost has less to do with adding air than in distributing composting bacteria from ground level to the rest of the pile. Adding occasional sprinkles of garden soil will accomplish the same goals with much less effort.
- Let your compost bins go thirsty or the composting bacteria will go dormant. If your bins have a solid lid, drill a dozen ½-inch (1.25 cm) holes so rain can drip through and keep composting bacteria in play.
- Get trashy by adding woody brush to a compost pile. It's just too dense for compost bacteria to break down between seasons. Save brush for the bonfire or let it break down naturally in nearby woods while creating habitat for chipmunks and food for mushrooms.
- Forget to balance "green" nitrogen-rich food scraps and garden scraps with carbon-rich "brown" materials like shredded tree leaves or coffee chaff from your local organic coffee roaster. I store chaff or shredded leaves in a scavenged garbage can next to my two compost bins. A short-handled shovel stored in the can makes it easy to top off "green" materials with plentiful "browns."

VDS: Vermin Denial Syndrome

I can't tell you how many times I've been given a tour of a gardener's compost setup and been assured they don't have rats or other vermin:

Me: "You happy with your open compost bin?"

Them: "Oh, yes! I dump kitchen scraps on top of the pile, and they have decomposed to nothing overnight."

(I kid you not, that's what they told me with a straight face.)

Me: "You're sure it's not being eaten by rats? Normally, it takes weeks or months for food scraps to turn into compost, which is different from just disappearing overnight."

Them: "Nope, never seen any rats."

Me: "What do you think made this fist-size tunnel under your compost pile?"

Them: "Huh. Never noticed that before. Maybe it's rabbits?"

Me: "Rabbits are mostly out during the day. Rats are mostly out at night. Have you ever seen any rabbits?"

Them: "Nope. Hey, come look at my fig tree."

WHAT ARE VERMIN?

YMMV here again, but if you are adding kitchen scraps to your leaky compost bin—or, if you leave cat food or dog food outdoors—you could be attracting a wide range of undesirable omnivores:

- Mice
- Possums
- Raccoons
- Rats
- Other critters depending on your location

I'm a fan of animals, generally. I've worked with a variety of domestic animals—cattle, pigs, chickens—and a variety of wild animals—sea turtles, rhesus monkeys in the wild, alligators—but when any kind of animal interferes with our little urban homestead, I figure they have sacrificed my forbearance and their lives are potentially forfeit. But better to avoid offering them accessible food and shelter in the compost bins in the first place.

Any of these animals could be carriers of rabies, could dig up your gardens, eat your eggs, vegetables, fruits, and pet food, tip over your trash and recycling containers, chew through cables and hoses under the hood of your vehicle, tangle with your cats, dogs, hens, rabbits, and any of them may get a notion to move into your cozy home. Ugh.

Composting *in* Stages

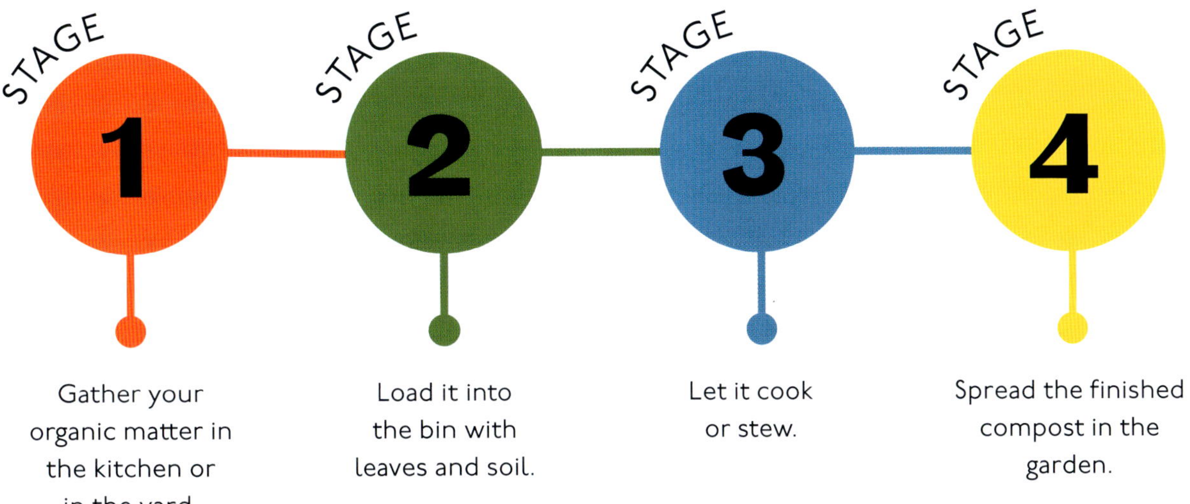

Gather your organic matter in the kitchen or in the yard.

Load it into the bin with leaves and soil.

Let it cook or stew.

Spread the finished compost in the garden.

Composting from Thirty-Thousand Feet

Let's look at composting by starting with the global perspective, shall we? The view from thirty thousand feet—or better yet, from the view of Earth in outer space—shows us one very good reason to compost your kitchen scraps rather than sending them to the dump. Our climate is being disrupted by excessive amounts of carbon dioxide (CO_2) and methane (CH_4) getting into the atmosphere and raising temperatures. And we feel that in terms of changes in our expectations of the weather. These rising temperatures stir things up. So, whatever weather behavior you grew up with, you won't get that anymore.

How do our humble kitchen scraps fit into that story? There is only about one-tenth as much methane as there is carbon dioxide in the atmosphere. But that smaller amount of methane has twenty to thirty times the impact of all that carbon dioxide.

One of the biggest contributions that a household makes to producing methane is the scrap foods and other bits of organic matter that go to the landfill. There, underground and deprived of oxygen, organic matter—kitchen scraps, garden scraps, wood, etc.—releases methane. If any of this organic waste were, instead, composted, none of that powerful methane would be produced.

If you're a gardener, instead of turning your kitchen waste into methane, you could be turning those scraps into garden vegetables or even eggs, if you—or a neighbor—have backyard chickens. So, let's get your kitchen scraps on the path to boosting your vegetable beds instead of disrupting our weather.

Infrastructure Tips

"Infra" means below. So, infrastructure is whatever tools or practices underlie your compost system. We'll start in the kitchen.

On the counter I have two tubs designed to hold kitchen scraps. Each has a lid and a handle. One is for scraps the compost bins like to "eat" and the other is for scraps our backyard chickens like to eat. If you don't have hens, you may want to have two tubs anyway. If you have a neighbor with chickens, they'll appreciate spending less of their money on bagged feed and will likely reward you with eggs that are fresher—and have darker yolks—than anything you can buy at the store. Win, win!

Because some leftover food is wet, I put a folded paper towel in the bottom of each tub so scraps don't stick when I want to dump them. And when take-out food comes with an excess of napkins, I stash them for that purpose too.

KEEPING STORED KITCHEN SCRAPS FROM STINKING UP THE KITCHEN

I keep our bins on the counter, near the trash can. That way, plates can be scraped into the containers and any food scraps that don't go into the compost—meat, bones, dairy products—can go in the trash.

Unrefrigerated food can last in the containers on the counter for several days before bacteria get to them and raise a stink. Before refrigeration was an option, families used to keep leftover food in a cabinet until it could be incorporated into another meal. We dump our kitchen scraps into our compost bins two or three times a week, so smell is never an issue. For people who want to go longer between trips to the compost, you could store kitchen scraps in the refrigerator.

Three Categories for Food Scraps

1. Food scraps from your kitchen—fruits, vegetables, grains, mushrooms at any stage of raw, cooked, or past their prime—can all go into your compost bins.

2. "Hen-able" scraps include almost anything your family eats **including** meat, seafood, dairy, and cooking oils. If you don't have chickens, I bet you won't have to look far for a neighbor who'll except your hen-able scraps from the kitchen.

   For example, I save bacon fat—some gets used to cook vegetables and mushrooms in our kitchen, as for the rest, the chickens act like its pudding once it's solidified in the fridge. And as carnivores, they will pick bones clean. Shrimp shells are like potato chips for them.

The things hens would not and should not eat include avocado skins and fruit, green potato skins, uncooked beans, and citrus. I advocate taking those long lists online of "things chickens shouldn't eat" with giant grains of salt. For instance, some people are concerned about seeds in apple cores because they contain a chemical that, in the stomach, *might* become cyanide. But studies show that a person—or a hen—would have to eat a 55-gallon (208.19 L) drum of seeds to have any negative effect. Large quantities of apple pomace—apple waste composed of seeds, stems, and skins—have been shown by studies to be a good food for all livestock.

What's more, there's also arsenic in many soils and that gets into almost all our food too. But it's at such a low level that our bodies can easily and safely deal with it. Ditto for the foods on these lists about what not to feed your hens. Urban henkeepers who haven't read these lists are not sickening or killing their hens willy-nilly. And chickens in the wild are not dropping dead from eating poisonous plants. They can smell and taste the difference, and you simply do not have to worry about that.

3 Kitchen paper products are another way to feed your compost bin and keep it happy, including with napkins, paper towels, tea bags, and coffee filters. Newspaper, printer paper, cardboard, parchment paper, seed envelopes, and paper bags would also compost in the very long run. But in the short run, when you put your compost in the garden, they just look like scraps of trash, so I forgo them.

Some people are reasonably concerned that their kitchen paper products may contain the antibacterial product formaldehyde, which, surprisingly, is made of carbon, hydrogen, and oxygen. But a study about composting plywood scraps, saturated with more formaldehyde than paper products would contain, found that even cool composting temperatures degraded formaldehyde to very safe levels. After all, it's just composed of carbon, hydrogen, and oxygen (a.k.a. CH_2O).

A real concern would be the microplastics in tea bags, but many of the popular brands don't use plastic. In which case, the tea, bag, string, label, and even the staple, if there is one, are safe to add to your compost bin.

LET'S *do a* DEEPER DIVE

Compost Science for Gardeners: Simple Methods for Nutrient Rich Soil by Robert Pavlis. The author maintains a 6-acre (2.42 ha) botanical garden. Pavlis is a myth buster and demystifier of compost science. The guy favors plain English and actual science.

CHAPTER 11

Growing Tomatoes for Chefs

I WAS TWENTY-FIVE YEARS OLD AND HAD NEVER SOLD ANYTHING TO A CHEF BEFORE, BUT I HAD BEEN DOING PLENTY OF SCHMOOZING AT A FARMERS' MARKET TO SEE WHAT THE OTHER ORGANIC TOMATO FARMERS WERE PROVIDING AND CHARGING. I'D ALSO BEEN CHATTING THEM UP TO FIND OUT ABOUT THEIR WHOLESALE PRICES AND PRACTICES.

When I knocked on the backdoor to the kitchen at the highly regarded Anotherthyme restaurant, I was nervous, but prepared. Suman, the chef, opened the door, releasing the smell of cooking onions and spices. She looked more closely at my heavy box of tomatoes than at me.

"Hi, I'm Frank, and I'm part of a cooperative farm that grew these organic tomatoes."

"I already bought a box of tomatoes earlier today," she said, nodding at the kitchen table behind her that held a 25-pound (11.33 kg) box (about half a bushel). But she couldn't resist picking up one of our tomatoes and looking closely at it.

"How much did you pay for those?" I asked, noticing how they were of random sizes, had cracks, bruises, green shoulders.

"Ten dollars. How much for yours?" (These are 1985 prices.)

Thinking fast, I said, "I can let you have this box for fifteen."

Shocked: "Why so much more?"

I nodded at my box and then nodded at the other box and said, "Look at them."

Her head swiveled from box to box. I held my breath.

She said, "I can use the others for soup. Let me get you some cash."

We were a bunch of tomato-growing novices, but we were getting the highest wholesale price for local, organic tomatoes.

For the next two summers, I sold our fresh, 1-pound (454 g), organic tomatoes not only to Suman at Anotherthyme, but also to the chef at the collective restaurant Somethyme, to the local natural grocery store Wellspring—soon to be bought by the Whole Foods chain—and to our tiny food co-op owned by a couple hundred households. The co-op's coolers and shelves were crammed inside what had been a rundown, two-bedroom home with an enclosed front porch, wooden siding with peeling paint, and roof shingles begging to be replaced. The legal name for the little co-op was "Intergalactic Food Conspiracy #1." There were dreams of expansion.

I had joined this little group of wanna-be farmers thinking that *I* would learn something from *them*. Turned out that my experience as an integrated pest management scout for conventional tomato farmers and a couple summers of growing my own tiny organic garden—seven raised beds just big enough to be confused with burial plots—made me the most experienced member of our little intergalactic farm conspiracy.

I proposed to my colleagues—including the co-op member who provided the land—that we combine the best of both farming worlds. We used organic techniques like:

- Spreading chicken manure from a neighboring farm for fertility and organic matter
- Using only organic pesticides, like Bt, for caterpillars
- Growing in raised beds for better drainage
- Starting thousands of plants from seed in fresh potting soil to avoid bringing tomato diseases from commercial greenhouses onto the farm

We combined those techniques with modern conventional farming practices that most organic farmers weren't aware of at the time:

- Rather than sprinklers, we used drip lines down each row to keep the tomato leaves dry and disease-free
- Black plastic mulch to suppress weeds, warm the soil in spring, keep the sun from drying out the soil, and to keep the rain from bouncing bad microbes onto the lower leaves
- Stakes and twine arranged in a Florida weave pattern—what I call a cat's cradle (see p. 86) to raise the fruit of our determinate tomatoes off the ground and make them cleaner and easier to pick

We were all ambivalent about the black plastic mulch, which would have to be pulled up and trashed at the end of each season. Happily now, farmers and gardeners can use compostable mulch that can be plowed in to decompose naturally.

By using this combination, we were able to grow and harvest pristine tomatoes that got us a top price, a great reputation, and a twice a week sellout of ripe tomatoes. And forty years later, I'm still acquainted with Suman, and we still get to enjoy her delicious Indian cooking.

Key Lessons for Selling to Chefs

If you want to ramp up from a backyard gardener to a market gardener, start by taking these steps:

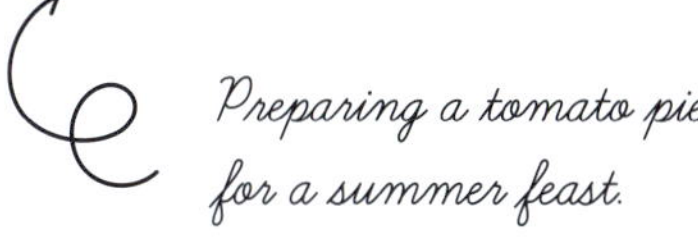

Preparing a tomato pie for a summer feast.

- Schmooze with fellow growers to get a sense of the lay of the land: prices, quantities, frequency of buying, personalities of the chefs or their buyers, and whether they've ever had a check bounce and what they did about it. Yes, that can happen. Not to us, fortunately. But you don't want to be caught flat-footed if it happens to you.
- Make a list of higher-end restaurants that might value locally grown produce and be willing and able to pay a fair price.
- Choose your most personable worker—maybe that's you—to establish relationships with chefs, buyers, kitchen workers. Anyone would rather do business with someone they *like*.
- Touch base with prospective chefs, ideally in person—not to get any commitment—to establish that they're willing to take a look at your produce when it comes in. And to show yourself as a trustworthy, reliable, likable person who communicates well.
- When your tomatoes are ripe, be very choosy. You are establishing your reputation with that first box. Some people think "organic" means "not beautiful." But there's no reason for that to be true, except in the case of low-skill growers. Any fruit that isn't beautiful, keep it for cooking, canning, or giving away. You want a reputation as the source for gorgeous tomatoes.
- Choose a good time to show up: meaning, not during a busy mealtime. Come by when it's quieter, between breakfast and lunch or between lunch and dinner. If you make a sale, ask how often they want you to come by and how much to bring. Use their buying schedule to determine your picking schedule.
- Show up with two minds: In one mind, you're confident of success. You're proud of what you've grown and not desperate for a sale. That's the kind of person a buyer wants to do business with. But in that second mind know they may turn you down. So, don't act deflated. If you do, the chef may feel like you're trying to make them feel guilty. And they won't be happy to see you in the future. Remember: Everyone wants to do business with someone they *like*. Also, everyone in business has been turned down. Be ready for it, stay cheerful, thank them for their time. Perhaps point

out a few of the biggest tomatoes and say, "Can I just give you enough of these for family meal?" That is the term for the in-house meal made from leftovers and scraps. They may accept or politely decline, but they'll appreciate your good cheer and your generosity. And, if they accept your gift and the kitchen staff remarks on the quality and flavor of your tomatoes, you may have just assured yourself of a future sale and a long-term business relationship.

- Be ready to negotiate on price. Have a figure in mind for what you think is a fair price. They may accept it. Or they may offer you less. So, know what your bottom dollar is. With Suman, I asked for $15, but would have accepted $12 or even the going rate of $10 per box if it came to that. We were just starting out and any sale would have been a win. If the going wholesale price doesn't cover your costs, look into selling at retail price at the farmers' market, even though that entails more time and more overhead than ripping through town, stopping at a few restaurants, and emptying your truck pretty quickly. Depending on your location—in a food desert far from grocery stores—you could get a retail price from honor system sales at a charming stand you set up for local drive-by customers. You may also want to explore "value-added" products. If you have a stellar recipe and a food-safety-approved kitchen, there may be a better price and a better market for roasted tomatoes, tomato sauce, salsa, and other products. The important thing is, do not be deterred. Explore options for selling your tomatoes—and other produce—in a fashion and at a price that make it worth your while. The purpose of a garden—even a market garden—is to make the gardener happy.

INFRASTRUCTURE

You want to up your game on growing tomatoes (and probably other produce too). What will you need to buy or build that you don't already have?

- A way to water without breaking the bank
- A place to store tools, supplies, and harvests that's out of the weather and safe from theft and vermin
- The rule for tools: better to have it and not need it than to need it and not have it
- A very quick primer on financial management for the small-scale self-employed grower

The details for building these many structures is beyond the scope of one book. So, I'm offering considerations that few novices are aware of and that I've seen many professionals overlook.

Speaking of pros, don't be shy about cultivating a mentor or reliable colleagues who can advise you based on their experience with these issues. My belief is that to learn about any subject, one must get a handle on the five to ten critical elements, and then get familiar with fifty to one hundred little nuances specific to your conditions that no book or video can adequately address. Get good at asking for help, feedback, and advice. And if—or rather, when—something doesn't work out as well as you hoped, don't keep a death grip on your original idea. Learn as you go. Own your goofs. Don't beat up on yourself. Take the lesson and move on. You're learning and getting to be outdoors, using your body, growing crops that make people happy. You're winning!

TOMATOES ARE MOSTLY WATER—THAT YOU MUST PROVIDE

As a backyard—or front yard—gardener you probably use city water from a spigot on your house to irrigate your tomatoes. Your municipality or some other entity has drawn that water from a river or reservoir, let the silt settle out, and treated it to remove pathogens. Then, they send you a monthly bill. The price of water in very dry regions will be pretty high. In high-rainfall areas, it will be much lower and may be your smallest monthly utility bill. Perhaps you reduce that bill with rain barrels or by building the rain box I detail on page 150. Either way, if you're going to grow enough produce, like tomatoes, to sell, that monthly bill will become a huge burden to your profitability. Eliminate that bill by hiring a pro to install a well to tap groundwater. This may be an option even on a modest-size lot in town, as long as the well-drilling rig can access your property. You'll pay a sizeable bill for installation, but you'll save money in the long run. And if you're lucky, the water—and water pressure—may be good enough to supply water for drinking and washing in your house. Be sure to plan on a reliable filtration system so your irrigation doesn't clog up.

Water from a pond can also be used for irrigation. You'll still need a reliable pump and filter system. Get some training on maintaining that equipment as anything with moving parts needs periodic attention. Better that it's you doing that work than an expensive mechanic.

FARMS HAVE BARNS AND SHEDS FOR A REASON

I love nature, but water is insidious, wind is indiscriminate, and the UV radiation in sunlight wants to damage almost everything it touches. Need I mention hail and snow? If your site doesn't already have a garage for storing tools and supplies, you'll need a well-roofed structure to play that protective role. Here are a few considerations:

- Any structure that holds anything edible—like organic fertilizer, chicken feed, your harvest, etc.—will need a concrete floor and tight walls and doors to keep out rats. A dirt floor is an invitation to vermin.
- The roof can be more of a DIY project by using what's called "rolled roofing." It resembles a heavy 3-foot (91.44 cm)-long paper towel roll, but it's made of the same material that much smaller roof shingles are made of. It costs much less and installs in much less time than shingles and lasts just as long.

If you aim to make money from your market garden, don't skimp on quality tools. But that doesn't mean you can't save money on tools.

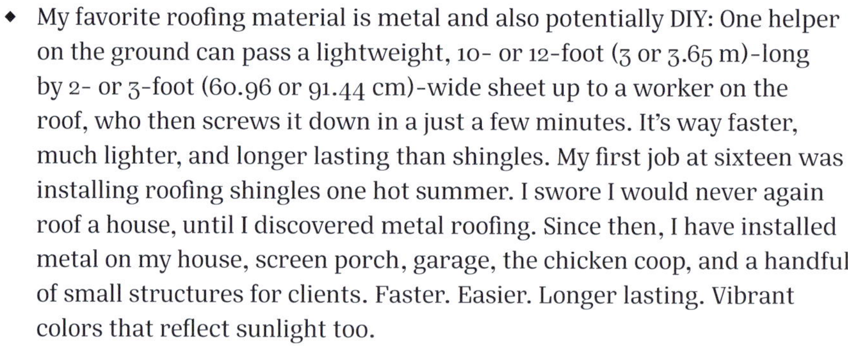

- My favorite roofing material is metal and also potentially DIY: One helper on the ground can pass a lightweight, 10- or 12-foot (3 or 3.65 m)-long by 2- or 3-foot (60.96 or 91.44 cm)-wide sheet up to a worker on the roof, who then screws it down in a just a few minutes. It's way faster, much lighter, and longer lasting than shingles. My first job at sixteen was installing roofing shingles one hot summer. I swore I would never again roof a house, until I discovered metal roofing. Since then, I have installed metal on my house, screen porch, garage, the chicken coop, and a handful of small structures for clients. Faster. Easier. Longer lasting. Vibrant colors that reflect sunlight too.

- If your location isn't susceptible to theft and you're not storing anything edible, you can build what's called a pole barn to store a tractor, supplies, and tools out of the rain. They have no walls or expensive windows or doors. The four or more poles support either a shed roof or a gabled roof—which is like two shed roofs rising to kiss each other.
- Hoop houses are now the classic structure for keeping rain off tomatoes in a moderate- or high-rainfall climate. Because they are often made of flexible plastic pipe covered by plastic sheeting, they don't stand up well in high-wind regions. You may need sheets of shade cloth to cover them for the hottest part of the season. You'll need electricity to run fans that blow out the hot air. The lower sides need to be raised or lowered depending on temperature.
- Greenhouses for starting seeds can be nice, if expensive, but sometimes they overpromise and underdeliver. If you *mis*manage heat, sun, and moisture on just one day a year, starts can be ruined.

FRANK'S RULES FOR TOOLS

It's better to have it and not need it than to need it and not have it.

Do not buy cheap homeowner-grade power tools. When companies design a homeowner-grade tool, essentially what they do is remove some expensive, long-lasting metal parts from a professional-grade tool and replace them with cheap, short-lived plastic parts. Buy professional-grade power tools.

Check out yard sales—and what we call junk-tique shops—for good prices. Like this shovel that I picked up for $10.

Ditto for professional-grade *hand* tools. But you don't have to buy them new. My wife and her mother used to go "bumming around" to what we jokingly call thrift-tique stores and junk-tique shops. And you can do the same. Work your way to the back, and that's often where you'll find a stash of old but good condition, and sometimes even handsome, hand tools for tomato gardening, farming, carpentry, stonework, and mechanical work. And they'll be just a few bucks each. The money you save will buy you a nice lunch on your bumming-around adventure. And even if you don't find something to buy, these places are a wealth of amusement and delight.

Tractors. Ask yourself—and your mentors and colleagues—"Do I really *need* a tractor at this scale?" In many cases the answer will be "no." Depending on how big your planting area is—and on how strong you want to become—your beds can be aerated with a garden fork or a U-bar or a rototiller or what's called a "walk-behind" tractor. If you conclude you do need a tractor, consider saving your too-scarce startup money by buying one that's used, since you probably won't be using it every day. I worked for two brothers with a woodworking shop and a cattle ranch. They bought a 40 h.p. used tractor with a front-end bucket. It was fifteen years old, with only one owner. So, it was in good condition, but it wasn't the most reliable brand. They figured that since they didn't need it every day, they could afford the occasional bit of downtime to take care of repairs on their schedule.

MONEY MAKES THE WORLD—AND YOUR MARKET GARDEN—GO 'ROUND.

Here's the skinny: Most market gardens and small farmers have off-farm income. It helps your effort a lot if your spouse has a straight job that pays well enough to keep your family in the black until your sales catch up and surpass your expenses. Or, if you have a flexible part-time gig that pays decently. Or, if you grew up working class and have mastered the skills for frugal living. On the other hand, perhaps you had the good sense to choose wealthy parents. If so, congratulations!

Don't pay more taxes than you have to. By that I mean get a separate credit card and separate checking account for your business. That way, all your business income goes into your business account, not your personal account. All your business expenses—tools, supplies, books, classes, work clothes, etc.—get paid through your business credit card and business checking account. Then, a couple times a month, pay yourself a "draw" by transferring money from your business account to your personal account. Those draws, added up, will be the basis for your annual income, on which you pay state, federal, and social security taxes. You don't pay taxes on business expenses. As someone once told me, "a small business is one of the best tax shelters there is."

Buy or rent tools? I own a garden design/plant/build business, but I don't own a rototiller. I don't own a Bobcat. Or a trailer. I need those tools infrequently. So, I rent when I need them, run them hard, then return them dirty to the rental business where they clean them and maintain them—until the next time I need them. I also don't own a dump truck. On occasions when I need a quantity of materials, I call a vendor. My time has better uses then driving around the countryside picking up and dropping off materials. You will want to own, maintain, and store as few pieces of heavy equipment as you can. The payments add up. And to pull out a big tool when you need it and have it grind to a halt for some mechanical problem is one of the most dispiriting experiences in life. In a lot of cases, it will make sense for you to rent or hire out the work.

Naming your business is a financial consideration. Why? Because just as people want to do business with someone they like, they also want to buy from a business that clearly offers exactly what they want. For example, when I started my garden business in 1992, I lived in a near-downtown neighborhood with older houses with front porches and small yards. When chatting with neighbors, it was clear that most homeowners in that demographic wanted a cottage garden. So, I called my business Cottage Garden Landscaping. People knew right away what they were getting. If you merely look around at many of the business names that cross your path in a day, you will likely find yourself wondering what they offer, as many business names are too clever by half. So, if I were starting a market garden, I might consider business names like "Always Fresh Organic Farm," or something more playful like "Ripe on Time Gardens." The more that customers *like* you and your helpers, and the more that your business name *resonates* with them, the more of their money they will throw at you.

PICK UP THE RIGHT PICKUP TRUCK

If you're going to be serious about growing lots of vegetables, whether for fun or to sell (or both), then you need to be serious about having the right vehicle. Over the years, I've bought five, ten-year old, small or mid-sized, used pickup trucks and then sold them when they reached the ripe old age of twenty:

- a Datsun long bed
- a Ford Ranger
- a Chevrolet S-10 (I normally wouldn't buy such a small truck, but when the retired guy who'd only used it for gardening chores lifted the hood, the involuntary thought in my head was, "I could eat off this engine." He had taken perfect care of it and the mileage was ridiculously low. He wouldn't dicker on the price. I bought it on the spot. It served me very well for ten years)
- a Nissan Frontier with a king cab and a tailgate extender
- and now another Nissan Frontier with a king cab (and I need to buy it a tailgate extender too. I'm a fan.)

There are lots of reliable used vehicles out there. But how do you tell them from the lemons (nothing against lemons per se). Here are some tips that might keep you out of a ditch with your used vehicle buying.

With modern websites for finding used cars, it's easier to search. I strongly recommend looking for these qualities:

- A mid-size truck with a king cab, so there's room behind the seat to store tools and supplies and other things that can't get wet. Like children and dogs.
- If it's an option, get a long bed rather than a short one. You can make either bed longer, by leaving the tail gate down and buying what's called a tailgate extender for under $200. These—mostly aluminum and plastic accessories—resemble a curved piece of fencing that rests on the outer edge of the tailgate and meets the quarter panels on either side, without blocking the rear lights. These extenders are easily removeable, let air flow through—very slightly improving gas mileage—and are sturdy enough to support a load of lumber that's longer than the bed and tailgate.
- Four-cylinder engines are powerful enough. I'd say buy six cylinders max, but even that isn't necessary for most professional or amateur gardeners. Trucks with eight-cylinder motors drink lots of gasoline and cost a lot of money to buy and maintain. And any load of materials that's too big or heavy for your mid-size pickup to carry, are much easier, cheaper, and less time-consuming to just have them delivered to your site. Any delivery fees will cost you much less than the upfront cost and long-term fuel and maintenance expense of an 8-cylinder pickup truck. Plus, you can be doing other, more high value work, in the time that a delivery truck is loading and getting to your site.
- Two-wheel drive is plenty. Only get a 4WD if you are dealing with rough terrain, like washed out roads. The more complicated a machine is, the more likely you are to have mechanical problems and the expenses that go with that.
- About ten years old is the sweet spot for me and the market I'm in. Your price points may vary in your region. But at whatever age, I recommend only buying a truck if it meets the conditions below.
- Use an online service for used vehicles to find one that has had no accidents at all. A damaged truck may have been "repaired," but there may be problems that aren't visible. For example, sometimes in accidents, the frame of the vehicle gets slightly bent. It won't be noticeable as you test drive it. But over time, the tires won't stay in alignment, the tread will wear unevenly, and you will be replacing them much too often.

When my truck was over twenty years old, my helper John bought it from me. That way, I get to still see my paint scheme: monarch butterflies and ferns.

- Buy a used vehicle that has had only one owner. That way you know that it was a reliable vehicle out of the factory. And you also can reasonably surmise that the owner took good care of it.
- But just to be sure, work out an arrangement with the present owner to allow you to take it to a mechanic for a "pre-purchase inspection" at your expense. This is a common practice that should take the mechanic about an hour or two to look for any obvious flaws. You can ask them to take more time to look for particular problems. Either way, this is always worth the investment, especially if they find a flaw that would be expensive to fix.
- Last bit of purchase advice comes from my father. He and his brothers had to drop out of school after the eight grade to work on their family farm in South Carolina. So, they knew a lot about buying vehicles and keeping them running. As he helped me pick out my first truck—a Datsun long bed with a four-cylinder engine—he told me, "don't fall in love with the first truck you look at. In fact, don't fall in love with any of them. You want to make a good choice. Not an exciting one." Once you've bought a great truck that cranks up every time and doesn't disappoint you, you will fall in love with it. And you'll feel wistful when you sell it. Thanks, Dad.

- Okay, this is really the last bit of advice. After I bought each ten-year-old truck, I sold my twenty-year-old truck to an individual. If I had sold the older truck to a dealer, they would have offered me a much lower, wholesale price. But that's roughly half as much as I get by showing it on weekends and after work to interested parties and getting the higher retail price. It requires more of my time, but the difference in price is well worth it.

And that retail price for a twenty-year-old truck gets me about half the money I need to buy its ten-year-old replacement. Have fun shopping.

LET'S *do a* DEEPER DIVE

The first thing I recommend you do before ramping up to market-garden status is buy these two books that helped me get a handle on successful self-employment:

Paul Hawken's book *Growing a Business* will help you get your business-owner head space right. In his twenties, he started Erewhon, the original organic grocery store, and he also started Smith & Hawkin, the top mail-order garden tool company.

Small Time Operator, by Bernard Kamoroff, CPA.

Both these people write very well about topics that could be painfully boring. But you're motivated and these authors are on top of their game. *The New Organic Grower: A Master's Manual of Tools and Techniques for the Home and Market Gardener* by Eliot Coleman is our bible.

About the Author

By training, **Frank Hyman** is a horticulturist, but by nature, he is a 'how-to culturist.' Meaning that he likes to figure out how-to make things more beautiful, more sustainable and more efficient. He spent his twenties as a very successful college dropout. After his sophomore year, he took a gap year that lasted eighty-four months. He was a working-class kid on a quest to find his callings. Along the way, he saved enough money to Jack-Kerouac around Europe for two months, but managed to stay for six months; drove work boats in fair and foul weather; worked on a barrier island with sea turtles; studied rhesus monkeys in the wild like Jane Goodall; worked as a truck driver, preschool teacher, bookseller, and archeologist; explored every phase of construction; and spent time as an organic tomato farmer. After finishing his double major in horticulture and design at one of the top hort schools on the planet, he started working for himself as a professional self-employed polymath, which affords him seven weeks paid vacation every year because he developed his callings into nine satisfying and AI-resistant professions:

Carpenter who builds arbors, decks, chicken coops, and treehouses up to 30 feet (9.14 m) above ground

Stonemason who learned from Irish, Welsh, and Spanish masons how to build patios and stone walls (without mortar) that will last five hundred years

Sculptor who wields a chain saw, stone chisel, or welding machine to create dramatic, outdoor art from copper, stone, trees, or scrap metal

Professional **gardener** who studied at one of the top horticulture schools on the planet

Award-winning **designer** creating gardens of all kinds

Political activist who cut off his ponytail to win a city council seat and authored the third living wage ordinance in the United States

Forager who sells mushrooms to restaurants, making as much as $200 per hour, and has also taught foraging to more than 1,500 chefs, arborists, organic farmers, cosplay Vikings, and members of the myco-curious public

Photographer who's sold more than two hundred photos to seven national magazines

Writer whose work has appeared in the *New York Times*, *Forbes.com*, the *Wall Street Journal*, and over a dozen national magazines and newspapers. He's written two popular books: *Hentopia* (do a search for his two-minute video with a surprise ending on YouTube) on backyard chicken-keeping and his best-selling *How to Forage for Mushrooms Without Dying: An Absolute Beginner's Guide*.

Acknowledgments

Thanks to all the lovely folks below who made this book possible and who made my story a delightful—and sometimes sweaty—reality.

The staff at Cool Springs Press: Jessica Walliser, Steve Roth, David Martinell, Mattie Wells, Gabrielle Bethancourt-Hughes, Mary Cassells, Tim Griffin, Kerri Landis, and everyone else who helped bring this book to life.

James Forrest White: Your example got me started on growing an organic vegetable garden at the age of twenty-two. You gave me that magazine full of articles about our garden guru, Alan Chadwick. But even that is not the best thing you did for me, aside from your friendship. It was that day in your dorm room when you handed me a mug of Irish coffee before that dreadful 8 a.m. class we hated. I moved too quickly and spilled some, and you said, "Be careful," and I said, "I was being careful," and you said, "Being careful means that one Is. Full. Of. Care." With those words, you changed my life, brother. Miss you.

Suman, you were the first chef I offered my organic tomatoes to. For the first time in my life I felt like I could accomplish important things that were valuable to other people.

Clemson Ag Extension agent **Cleve Saunders** and the hard-working tomato farmers of Saint Helena and Lady's Islands, South Carolina.

Models: Olga Fedorovna, Dyson Robinson, Zoe Shear, Echo Wilson, John Hawkins, and photography consultant Liz Nemeth.

North Carolina State University design profs: Tracy Traer, Will Hooker, Fernando Magallanes, Denis Wood, Curtis Brooks: thanks for the guidance and education.

All my North Carolina State University horticulture professors: Stu Warren, Paul Fantz, Bryce Lane, Mike Parker, Rachel McLaughlin, and Kim Powell.

My professor and mentor in horticulture and writing, the late **Dr. J. C. Raulston**. Thanks for helping me see the world from the perspective of plants. And for kicking my butt to start pitching stories to garden magazines.

Thanks to my **North Star, Chris Crochetière**.

Green tomato neighbor, Mrs. Calhoun. Thanks for exposing me to proper fried green tomatoes.

Co-op farm buddies: Thanks for putting your faith in me: Deck Stapleton, Ruffin Slater, Tom O'Conner, Mikel Taylor, Michael Meredith.

And last but not least, thanks to colleagues Shawna Coronado and Steven Biggs for providing ideas and encouragement on this book.

Index

R

S